A NEW ZEAR

DERIC LOVE

outskirts press

A New Zear
All Rights Reserved.
Copyright © 2023 Deric Love
v3.0

The opinions expressed in this manuscript are solely the opinions of the author and do not represent the opinions or thoughts of the publisher. The author has represented and warranted full ownership and/or legal right to publish all the materials in this book.

This book may not be reproduced, transmitted, or stored in whole or in part by any means, including graphic, electronic, or mechanical without the express written consent of the publisher except in the case of brief quotations embodied in critical articles and reviews.

Outskirts Press, Inc.
http://www.outskirtspress.com

ISBN: 978-1-9772-5363-7

Cover Photo © 2023 www.gettyimages.com. All rights reserved - used with permission.

Outskirts Press and the "OP" logo are trademarks belonging to Outskirts Press, Inc.

PRINTED IN THE UNITED STATES OF AMERICA

There is a GOD! Even if you do not believe that there is a GOD, you will believe that there is truly a GOD after reading this book. I have no prior education regarding Astronomy, or in any other fields that are related to studying the universe; moreover, I have never had a job in the field of Astronomy, e.g., work for NASA, or any other companies or governments, which study the universe. The knowledge, which is revealed in this book about the earth in its relationship with the universe, was giving to me via a vision and direct communication with GOD. If my conclusion is wrong, then I am hearing imaginary voices in my head, and I am delusional about hearing the voice of GOD; however, if you find that I am revealing the truth about the universe, which every civilizations, schools, governments et cetera have tried to comprehend but fail, then you must accept the fact that there is a GOD who revealed this to me.

CONTENTS

The Beginning of Time 1
A Day. 6
North, East, South & West 10
There Are Sixty Minutes in an Hour 11
Six Hours is a Frayation 13
The Great Discovery of Frayations 14
The Distance of the Earth Around The Sun 16
The Number 28. 22
The Discovery of a Tay. 37
The Days of Our Lives. 40
Throughout the Years. 42
What is a Deason. 51
A Zear is 1,460 Years 56
The Comyetion 63
Five Point Star 69
This is Wrong. 79

On Dead Man Time. 80
Different Time Zone 83
A New Math 84
Time is Moving 85
Biblical Facts 88
Notes 90
Endnotes 92

THE BEGINNING OF TIME

I AM INFATUATED with words. I know that articulation is important when speaking to others. However, I find that the meaning of words has value, too. The meaning of words creates thoughts in your mind, which open the mind to a new way of thinking, reasoning, and expressing yourself. One word (in particular), which I find very interesting, is time. I love how the Webster dictionary define time, which is, "duration regarded as belonging to the present life" … in other words, time measure the length of time that the present life continues and exists. This also include time to measure the length of time that something exists or continues in the present life. Therefore, time is very important to the present life. I ask you to keep this meaning in mind, while you are reading this book, because life is not measure by liter, gram, meter, volume, but time: in other words, we do not measure the duration of life, or the duration of the life of something by measuring it in distance, its weight, area et cetera, but by its position in time. The reason for the correct calendar system has remained a mystery, because we are measuring the distance of the earth in the solar system instead of observing its position in the solar system, e.g., time is indicated on a clock by observing the position of the hands on the clock instead of observing the distance of the hand from the beginning to the end. When you want to learn the distance of something, you measure its length;

however, when you want to learn the duration of the life of something, you measure it by comparing it to the position of time, right. So, follow me as we observe the position of time in the solar system, which has never been discover or discuss until now.

I would like to bring to the human race's attention that by the earth to be six hours from its original position after a year is not a random movement, but I have detected an arrangement of repeated/corresponding pattern between years and days. Therefore, instead of trying to start the years in the same position by adding a day to the calendar year on the fourth year, I recommend to just observe how the earth shift six hours to start the new year at a new position, and you will observe a pattern. If you do not follow this pattern, you will make an error in interpreting the purpose for the earth to start a new year in a different position than the previous year. Others have been able to figure out the pattern of seconds, minutes, hours, and days; however, they were not able to figure out the pattern with the years, because they lack an understanding of the function and purpose of six hours, which is label as a frayation in this book. We must agree that by not understanding why the years start in a new position, this is causing us to not fully understand the purpose of time. If there is a reason for the days to form four seasons, which affect our lives, then there is a reason for the years, since it forms the same pattern as the days; therefore, we are missing the reason for the years, and what influence that it has on us. Since years form a similar pattern as days and frayations, which bring about the season, and a new year, then let's observe the years more closely to observe its configuration of four seasons, and the newness that it brings, which I have done in this book. We will also observe six hours (that is, frayations) in this book, since they, too, are forming four seasons in a year as the days to form four seasons in a Tay. Here is just one example of these frayations, and years to form a pattern as the days. Notice how the first and last day of the year is on the same day; the years and frayation will begin and end likewise.

THE BEGINNING OF TIME

Many have tried to comprehend the solar system but have fail with each attempt. Some have come close to figuring out the system with the solar system, but this is not a game of horseshoe; therefore, closeness does not count. For no one want to know or ask what the time is almost, but we want to know the exact time. Time is very important to us, because we use time to meet with each other; time affects the earth, e.g., certain time of the year is spring, fall, winter, or summer. The nature of time affects the season of harvest. This process of time cleanses the earth, produce food for us et cetera. There is an old saying, "there is a time to die, a time to be born, a time to plant, a time to uproot (that is, a time for everything). What if we want to work together, have fun together, train together, or do whatever come to our mind together. We must set a time to do it, or we will miss each other. Time is the most fundamental element to organizing, preparing, determining, or beginning our affairs. Therefore, we cannot accomplish anything together without a time to do it together. Time helps eliminate the confusion of when something must occur. We use time to calculate when to shoot a target. Therefore, time is used to make sure something or someone meet at the appointed time. I can go on and on to prove how relevant time is to your daily living and to nature, but I think that I have given enough evidence to show its value.

The reason why people are confused and miss this pattern who study time, because everyone's attempts to solve this mystery is approach by analyzing the distance of the earth between the start of the year and the end of the year. However, time is not measure as you measure water, a line, the weight of something, the distance between two points et cetera ... you must use a different measurement to measure time. You tell time with a clock by the positions on the clock instead of measuring the distance between points, that is, these points are a position in time without extension. They figure out how to place the positions on a clock to indicate the time of day; nevertheless, no one has thought to create something like a clock to indicate the time of week, month, year, Tay et cetera, which this book is explaining to you. If we were to create a clock to indicate the entire system of time, that is, not just a clock that tell the time of day, hour, minute, and second, but also the year, Tay, deason, and Zear, this would be a clock in more detail about time. There would have to be 1,460 positions on the clock. Each position to represent a year. Every fourth position to indicate a Tay. Divide 1,460 years by 4, which is 365 years (that is, a deason). Each deason indicate the years to start in the area of one of the four seasons. This complete process is a Zear.

Julian Caesar and others were trying to figure out a calendar to cause the earth to start the new year in the same position. Nature would have to eliminate six hours for this theory to work. This mysterious point in time has caused other's fail attempt to solve this mystery to not be able to interpret it, or explain it until now, which this book will explain in full detail for the reason for the earth to not start a new year in its original position, but it will start a new year, which is six hours away from the previous year. Moreover, notice that there is 365 days in a year. For the earth to start the new year on the same day, it would have to be 364 days. This extra day is what causing the earth to start the new year on the next day, e.g., if the first day of the previous year was on a Monday, and the last day of the same year is on a Monday, then the first day on the following year is on a Tuesday. More on this topic in more detail, later in this book!

It would have been easy to determine the exact days in a month, and the exact months of a year, if it was 364 days in a year, because you can evenly divide the days with 364, and the whole world would agree to have the same calendar. Yet 365 days just threw a monkey wrench in the game, which just confuse the whole world. All have agreed that their calendar is not exact as the solar system, but as close as they can think of ... if they would just focus on the effect of six hours at the end of the year, they would figure out the exact pattern instead of coming close, which is what this book will reveal to you, that is, this book is revealing a calendar that is exactly as the solar system.

One mistake that these calendars make is that a year reflects a complete cycle of the earth in the solar system, which is crazy. It is impossible for the earth to align in its original position at the end of one year (that is, it is impossible for the earth to complete its process with nature in one year). The universe and earth are too big for such an event to take place in one year. One year is not even a degree of the time to reflect the earth to complete its cycle. It takes thousands of year for such an event to take place.

Another reason why no one has been able to completely discover the calendar year, that is, the pattern of the earth in alignment with the solar system, because they falsely assume that the earth should be in its original position at the end of the year. Why no one never thought to accept that maybe it is not supposed to be in its original position at the end of a year. Notice that if it is Monday, the earth will not be in the same position on Tuesday, right. Since the years are forming the same pattern as the days, then the year is not going to be in the same position at the start of a new year, e.g., if the new year is on a Monday, then the following year will start its new year on a Tuesday and so forth with the following years, right; moreover, the new year will start in a different position than the previous year. However, we look at the years as the same, since we are so determined to figure out a calendar year with the earth to start a new year in the same position. Since I have revealed evidence of the pattern with the years is like the pattern with the day, notice how we say, "what day is it" instead of looking at the days as the same day. We should look at the years as the same, e.g., say, "what year is it?"; Tuesyear, Wednesyear, Thursyear or whatever we decide to cause the seven years.

Notice that there is a consistent pattern with the seconds, minutes, hours, frayations, weeks, and days in a year; however, there is no pattern by creating a leap year; as a matter of fact, the creation of a leap year is disrupting (that is, destroying) the pattern with the years.

A DAY

SINCE THERE IS 24 hours in a day, the following must be noted: 24 hours is 360°, 12 hours is 180°, 6 hours is 90°, 3 hours is 45°, and 1 hour is 15°, which compute for the earth to turn 15° at the end of each hour (i.e., 60 minutes), which narrow down to the earth to turn five degrees in twenty minutes, and every four minutes reflects one degree.

I will begin revealing the greatest mystery of time, which is time; however, I will start by discussing some basic general knowledge, which will become very important to remember the following information about a day, that is, later in this book. Therefore, highlight, or put a bookmark right here to remember this page.

There are sixty second in a minute; therefore, 60 seconds is 360° of a minute, 30 seconds is 180° of a minute, 15 seconds is 90° of a minute, five second is 30° of a minute, and one second is 6° of a minute. There are also sixty minutes in an hour, right; therefore, again, 60 minutes are 360° of an hour, 30 minutes are 180° of an hour, 15 minutes are 90° of an hour, 5 minutes are 30° of an hour, 1 minute is 6° of an hour, and every ten seconds is 1° of an hour. Not only is six hours (that is, a frayation) is 90° of a day, which is a ¼ of a day, but notice how the day is divided to resemble the four seasons: the sun figuratively rises in the morning like the spring, set at its hottest point in the afternoon like the summer, the sun falls in the evening like the fall, and it rest at midnight like the winter. This will appear as a coincidence, but you will learn later in this book that this is not a coincidence but done with intention.

Moreover, since we are taught that the sun rise, and set, we view the sun as passing, that is, rising in the east, and setting in the west; however, there is one error with this, which is, it is impossible for the sun to rise and set, because the sun does not move. The earth is what is moving! Since the earth is moving instead of the sun, you can tell how the sun is in the east in the morning and in the west towards the evening, this is enough evidence to prove that the earth is moving counterclockwise instead of clockwise; in other words, the earth is turning towards the left.

We think, that at the end of the year, the earth is 6 hours behind (i.e., off) its course. So, it appears that time is moving backward. Therefore, in 4 years, which is a total of 1,460 days, the earth is 24 hours behind (i.e., off) its course. We are limiting our understanding of time by analyzing the pattern of time within a year, or four years. However, if we analyze time after a period of 1,460 years, or 10,220 years, we will better understand, and discover the pattern of the years. The reason why I chose 10,220 years, because after studying the pattern of the earth, I have learned/discover that the earth is in its original position and starts the year on the same day as the first year. Again, we think that time is moving backward, or is 6 hours away from its original position during the start of a new year; however, after analyzing time on a larger/broader scale, I have learned that time is moving ahead instead of falling behind, and the earth is shifting ahead, too. Yet we are taught that the earth is falling behind. You will learn in this book that this teaching is not correct, which is due to a misunderstanding of time. I will prove that the earth has not fallen behind, which we say, a day after four years, but it has advanced a day after four years, and it has advanced six hours after one year. Therefore, let us focus on six hours, which is label as a frayation in this book, to unravel this mystery with time. I bring this to your attention, because 6 hours is the focus point to understanding time. Notice that there is 360 minutes in 6 hours. Since six hours is a complete fundamental cycle to time, then each minute is a degree within the 360% of six hours. Therefore, one minute is one degree of six hours, and one minute reflects the earth to turn ¼ of a degree. I must stress, again, that a frayation (that, six hours) is fundamental to time, that is, the basic to understanding time. However, we think that 24 hours is the basic, which is a mistake. Since we are confused about the earth in a different position at the end of the year, which is six hours away from each other, this is what made me to focus on six hours. If we focus on six hours, then we can better understand that the earth is not six hours off its course at the end of a year, but it has completed its course to advance to the next position, which is six hours ahead from the previous year.

Each year starts six hours away from the previous year, and there are 1,460 positions for the earth to start a new year; in other words, there are 1,460 frayations (that is, six hours) in a year; moreover, since a new year will start six hours away from the previous year, then there are 1,460 positions for the earth to start a new year before it to start a new year in its original position. What a coincidence, or is it a coincidence, or a pattern? For us to really understand how time work, we must stop interfering with time, that is, stop adjusting time after each four years, and just observe as nature take its course, and you will learn how to figure out the patterns with the years. We will also understand why the position of a new year shift six hours at the beginning of a new year. For while we are trying to position the earth in its originals position after four years, this is moving our calendar system out of a parallel position with the function and position of the earth in according to the solar system. The earth reflects time; therefore, how we adjust time after four years, this reflects our attempt to make time to stand still in one position. Yet the earth is constantly advancing in position in the solar system, which reflects time to not set still, but constantly advancing through the years and with the years.

We are practically lost in time, because if time (that is, our calendar system) is supposed to reflects the location of the earth on its plane around the sun, then the exact position of the earth is lost by determining it by our calendar system. This is a fact, since everyone attempt to create a calendar system will admit that they do not fully understand the earth in relationship to the solar system. Therefore, our calendar system is not an exact representation of the earth in the solar system. We need to create a calendar system to reflect a way to be parallel with earth in a correct manner, or our interpretation of time will always be incorrect. This book reveals a calendar system to reflects exactly not only what second of a minute, minute of an hour, hour of a day, day of a week, week of a month, month of a year, but also year of a deason, Tay of a Zear, and Zear of a Comyetion.

NORTH, EAST, SOUTH & WEST

AFTER FOCUSING ON the earth, which is moving around the sun instead of the sun to move around the earth with a sunrise and sunset, I notice that the earth is like a compass within itself to explain its location in the solar system, which is interpreted as not only the time of day, but also the time of year, Tay, Deason, Zear, Comyetion et cetera.

THERE ARE SIXTY MINUTES IN AN HOUR

THERE ARE ALSO 360 minutes in six hours, which is a **frayation**[1]. A frayation is a new word, which is used to label a new important discovery about six hours, which is discussed in further detail in this book (but later). You will learn in this book about a frayation, which is the foundation to building the four seasons, and the major positions (that is, points) in time of the start, end and completion of each season, day, week, month, year, Tay, Deason, Zear et cetera. Since there are 60 minutes in an hour, there are 1,440 minutes in a day, which come to 525,600 minutes in a year. Since there is 360° in a complete circle, then every 1,460 minutes is 1°; in other words, since it takes 525,600 minutes for the earth to complete its orbit in a year, and if you divide 525,600 minutes by 360°, you get 1,460 minutes. Therefore, it takes 1,460 minutes for the earth to move 1° on its course around the sun … this is interesting, because it is 1,460 frayations around the sun; there is also 1,460 days in four years (the significance of these days is discussed lately); it takes the completion of 1,460 years before the earth to start a new year in its original position. Since it takes 1,460 minutes for the earth to move 1°, then it takes 365 minutes for the earth to move ¼ of a degree. Notice that 365 frayations indicate the earth to complete ¼

A NEW ZEAR

of its orbit around the sun to complete one season; it takes 365 days for the earth to complete one year (that is, season), or ¼ of a Tay (more on a Tay lately), 365 years for the earth to also complete a season, which is identified as a deason in this book. 1,460 minutes are equal to one day and 20 minutes. Please, be patient! I am about to reveal a beautiful pattern with this point, which is known as the five-point star.

SIX HOURS IS A FRAYATION[2]

I LABEL "A six hour span" as a **frayation**, because six hours is the foundation of time, that is, everything is built, connected, starts, ends, or related to six hours in some type of way. Since there is 360 minutes in a frayation, each minute is considered as 1° of a frayation. When a new year begins, the earth is exactly six hours away from the previous year. Six hours is also ¼ of a day. In other words, there are 24 hours in a day, and there is a 24-hour time zone (that is, there are 24 hourly divisions on earth, right); therefore, the part of the earth that is facing the sun on the new year is 90° away from the previous year, which is 360° after four years.

- There are twenty-eight frayations in a week.
- One day can be divided into four frayations, and each frayation symbolize one of the four seasons, e.g., midnight symbolize winter, morning symbolize spring, noon symbolize summer, which is the hottest part of the day, and the evening symbolize fall.

THE GREAT DISCOVERY OF FRAYATIONS

365 FRAYATIONS IS ¼ of a year; therefore, there is 1,460 frayations in a year. Moreover, every position at the beginning of a day (this also includes dividing a day into 4 frayations), the beginning of every new year, and any other patterns that come to your awareness with time, and any other new patterns that I will bring to your awareness in this book always start and end in one of these positions of a frayation. If you observe ¼ of a year, which is 365 frayations, you will notice that it starts and ends at the same time of day, that is, if it is 6 am in the morning, 365 frayations away from here is 6 am in the morning, too. I am bringing this pattern to your attention, because if you follow it, you will discover how it relates to the four seasons, e.g., the first 365 frayations start and end in the morning, which is spring; the next 365 frayations will start and end at noon, which is summer; the next 365 frayations will start and end in the evening, which is fall; the last 365 frayations will start and end at midnight, which is winter. I am not making up something to follow, which Julian Caesar and others have done with their calendar system. I am revealing an actual pattern of time for you to examine for yourself. Therefore, do not take my words on anything that I say. Add these frayations up and calculate for

THE GREAT DISCOVERY OF FRAYATIONS

yourself by observing the beginning and end of each day, the beginning and end of each year, the beginning and end of each position within a day by dividing a day by six hours, and notice for yourself that every new positions start and end at the beginning of a frayation. If you come to the same result, then we can conclude that these frayations are the fundamental start and end of each day, week, month, year, Tay, deason, Zear et cetera instead of a day as the end and start to each pattern. If you just focus on the day, your timing will always be off, and overlapping another day, but if you just focus on the pattern of each frayation (that is, six hours) within a year, everything just seems to fall into place.

THE DISTANCE OF THE EARTH AROUND THE SUN

ONE MISTAKE THAT we make with time is the misinterpretation of 365.25 days for the earth to orbit the sun. This is somewhat true. It does take 365.25 days for the earth to reach its original position; however, you cannot measure time by space. Time is interpreted by position, e.g., you do not measure the distance between the beginning and end of a minute or hour, you observe the position on the clock to determine the time of day. Moreover, there is no such thing as a negative number with time, subtracting time, or zero time, e.g., nothing happens in zero time. Zero time means that it has not started. Measuring time is not like measuring space, or an object, e.g., what if I told you that I bought ten apples, and I eat 3 apples, then I ask, "how many apples are left"? You would subtract three from ten, and say that I have seven apples; however, if you were to time the amount of time for me to jog ten miles, then time me in the amount of time to jog backward for three miles, you would not subtract three miles from ten miles to get the result, because time does not go backward. No matter what happen in time, it is considered as going forward in time. Time never goes backward. There is also no position for zero in time. Notice how zero does not exist on your clock, watch, stopwatch, or anything

THE DISTANCE OF THE EARTH AROUND THE SUN

else that tell time, because, again, zero does not exist in time. So, when considering the position of time instead of the distance of time, I have discovered that there are 1,460 important positions in time. I figure that these positions are important, because every day, frayation, week, year, season, month et cetera always starts and ends in one of these positions; moreover, it never starts between these positions. This is how you can figure out these positions for yourself with my opinion on this matter. There are 24 hours in a day; consequently, there are 8,760 hours in 365 days of a year. Notice how the earth is six hours away from its original position at the end of a year. If you were to follow this pattern without including a leap year, it would take 1,460 years before the earth to start a new year in its original position. Therefore, this is one evidence of the earth to have 1,460 position to start a new year. Notice that the beginning of a year, ¼ of a year, ½ of a year, ¾ of a year, a whole year, a day, the seasons, and anything else with time occur at one of these positions, and it never occur between these positions. Therefore, these are the key points in time. Remember, there is no such thing as a zero position in time, right. Again, zero does not exist on your clock, stopwatch, or watch. We make the mistake to say that it takes 365.25 days for the earth to orbit the sun as a sign of completion, because, again, we are measuring the distance of the earth around the sun. Yet the completion is 365 days, e.g., let's look at the movement of the earth in terms of position instead of distance. The first position is the first day of the year at midnight, right. Well, you cannot count this position anymore, because it is the first position. Yet when we measure the earth to orbit the sun by distance, we count the first position twice by considering it as the first and last. When the earth reaches the last position, it is six hours away from the first position. In other words, 365 days is the amount of time for the earth to position itself at every position around the sun. The last position for the first year is also the beginning of the following year, which is six hours (that is, a frayation) away from the previous year. I can explain it in this matter. Notice how the hands on the clock are always moving, but the position is constantly in the same position. Time is really working like that with years. For example, the proof is in nature, because no matter how much you try to make the year to start in the same position, you just don't seem to be able to figure it out, right. However, you have figure out that a new year always start in a new position. You should accept this act of nature by observing the years to change to a new position. Years are forming a pattern like days, frayations, deasons, and other patterns of a Zear, which is discussed lately in this book. Therefore, we know that no days of a year is in the same position, but it is another year when a day is in its original position of the earth to orbit the sun. However, it is a different time of day. It takes four years, which is 1,460 days, before the earth to start a day at it original position at the same time of day. The years are forming the same pattern, because it also takes 1,460 years before a new year to start in its original position.

There is a logical reason for the years to change position. If the earth was in the same position on each year, this would cause an imbalance effect with the nature of the earth. The earth is changing to a new position at the end of the year to evenly cause the earth to have a balance nature, e.g., what if the earth was to set still. Half of the earth would probably be frozen, and the other half would probably be extremely hot. With the years to cause the earth to change to a new position at the beginning of a year, this causes the sun to shine on the earth in a different position than the previous year. It takes four years for the earth to change to cause the sun to shine on all parts of the earth, that is, whatever part of the earth is facing the sun at noon on a new year, when the earth reaches this position on the following year, a different part of the earth will face the sun, which is 90° away (that is, six hours away) from the previous year. This is one of many reasons for the new year to start in a different position. You will learn more reason as you read this book!

Everything about time to start and end the day, week, month, and

THE DISTANCE OF THE EARTH AROUND THE SUN

year will change position during the following year by six hours (that is, a frayation). Tays and Deasons will start and end in the same position during the following Zear, because they indicate the location of the seasons during a Zear. We know that the location of the seasons to never change position. A Zear never change position, because it is the totality of the complete path of the earth to orbit the sun in 1,460 years; needless to say, these frayations are in a permanent position, too, because they indicate each position for the start and end of a day, week, month, Year, Tay, Deason, Zear, and any other pattern with time. Think of a clock … the number on a clock never changes its position; however, the beginning of a new second, new minute, and new hours change position until it makes a 360° turn on the clock; moreover, the numbers on the clock indicate the position of a second, minute and hour to interpret the time. We have this education about time. I am furthering your education about time to understand the action of a day, week, month, year, Tay, deason, and Zear are like this behavior. These 1,460 frayations are the position on the clock to interpret what time of day, year, month, week, Tay, Deason, Zear, et cetera. The Days, years, Tays, and Deasons must move to each of these positions of the 1,460 frayations to complete its cycles. 1,460 days, which is a Tay, is forming the same pattern as 1,460 years, which is a Zear. The earth will never change to take a different course to orbit the sun. This book labels this course as a Zear. In other words, the earth changes position to orbit the sun; however, the path that the earth takes to orbit the sun, this will never change. I have established enough facts to bring a conclusion about the mystery of time.

Only nature arranged for the earth, stars, moon, et cetera to all align at their original position at its appropriate time, not when we want this event to take place after four years. When we try to adjust time to our understanding, it throws us out of sync of unknown things.

This is one evidence that the earth is not drifting backward at the beginning of the following year; on and to the contrary, the new

A NEW ZEAR

position of the earth at the beginning of the following year is a sign of the earth to reflects its advancement, because if the first day of the year is on a Monday, then the beginning of the following year is on a Tuesday. So, a year reflects the earth to make progress at the start of a new year.

While man is trying to make a new year to start in the same position, nature has time on a movement to a new position, which is seen in a pattern of advancing six hours at the start of a new year.

Go to your calendar! Notice that the first and last day of the year is on the same day, which cause the following year to fall on the next day. This is just another sign of time, which reflect the earth's movement, that the earth is not drifting after each year, but advance to the next day. This pattern is seen with the earth with the days. That is, whatever day it is, 365 days later, which is a new year from now, will fall on the next day. If you have looked at a calendar, and did the math on your own, you will see this to be true. And if you now follow this pattern for every 365 years, you will see the earth to advance likewise. Now for the finale! There are twenty-eight 365 years (that is, 28 deasons or seven Zears), which is a total of 10,220 years (this book label 10,220 years as a Comyetion[3]), before the earth to start a new year in same position, on the same day, and at the same time of day by nature (i.e., without man's interference with adjusting the calendar after four years). Notice with the 28 deasons (that is, seven Zears). This is like a month with four weeks, and 28 days. The same pattern seems to form after the end of a Comyetion[4]. Which is why I say that this is the completion. However, since 28 Deasons form this pattern, and 28 is never a sign of a completion of a pattern with anything else with time, but a stage, my guy feeling is telling me that this pattern must occur 13 time before the completion of time. In other words, it is a scientific fact that the same pattern seems to repeat itself after a Comyetion. So, there is no other pattern after this. However, my intuition is telling me that something else is happening after a Comyetion, e.g., the earth will start a new year

THE DISTANCE OF THE EARTH AROUND THE SUN

in its original position after 10,220 years (that is, a Comyetion), but what about other things in universe. I would love to know when the earth, stars, moon, other planets, and everything else all align at the same time at its original position. In other words, this book is covering facts about the earth (not every entire things in the universe, which is also a strong interest to me).

THE NUMBER 28

THERE ARE 28 seasons (that is, summer, winter, fall, and spring) in seven years. There are 28 frayations[5] in a week, which has the characteristics of seasons, e.g., if you divide a day into four divisions equally, you get six hours in each division. Midnight is the coldest part of the day like winter; morning is like spring with the sun to rise; the afternoon is the hottest part of the day like the summer; the evening is like fall with the sun to fall. This pattern with the day is no coincidence, but how the days, weeks, months, years, deasons, frayations et cetera are forming the same pattern with four seasons. Since a day has the characteristic of four seasons, then seven days, which is a week, is 28 seasons (there is also 28 frayations in a week). This book will reveal so many great discoveries with time, e.g., there are 28 days in a month. Notice how 28 days in a month form the same pattern as a week, which was just discussed. Each day has four seasons, which is 28 seasons in a week. With 28 frayations in a week, this will cause 4 weeks in a month.

There are always sixty seconds in a minute, then it starts a new minute after this. There are sixty minutes in an hour, then a new hour starts after this. There are 24 hours in a day, then a new day starts after this. There are seven days in a week, then a new week starts after this. There are 365 days in a year, then a new year starts after this. This is a consistent pattern with time. Yet there is not a consistent pattern with the months according to a calendar. One month has 28 days, some

THE NUMBER 28

months have 30 days, and other months have 31 days. The inconsistency is proof that this is not the correct calendar year. I do not say this in a negative way, because Julian Caesar admitted that he could not figure out the correct way to make a calendar. This is as close as he could get … We must admire his effort, because no one has gotten closer than Julian Caesar to figure out a calendar to reflect exactly the earth to orbit the sun until this publication.

This book will educate you about having 28 days in each month to not only have a consistent pattern, but the reason for having 28 days in each month. Each month will start on the same day of the week with 28 days, and 28 days reveal the days, weeks, months, years, frayations et cetera to establish the same pattern, and have the same function. Notice with our present calendar system! The seconds, minutes, hours, days, weeks, years, Tays, Deasons, and everything else except the months create a pattern like something else. This book reveals the month with a pattern as every other pattern with time.

Since every pattern are similar, notice the similarity of a week and a month with 28 days in a month. There are 52 weeks and four frayations in a year; thirteen weeks and a frayation (that is, six hours) in ¼ of a year. With 28 days in a month, this will cause 52 months and four days in four years, and 13 months and a day in a year. Keep in mind that it is not the days that are forming a pattern in a year, but frayations. Days form its complete pattern in four years. Keep also in mind that the months should have a consistency result as every other pattern with time. The reason why no one could figure out the correct number of days in a month, because we are under the impression that a year is the complete cycle of time; however, the years are like the seconds, minutes, hours, days, weeks, and months, which add up to build something else. This book will reveal how years are building up to form the same pattern as the days, frayations et cetera. I ask you to pay close attention to the pattern in the following illustration, which is the correct calendar system.

A NEW ZEAR

JANUARY 2023

(The year of Sunyear)

Sunday	Monday	Tuesday	Wednesday	Thursday	Friday	Saturday
1	2	3	4	5	6	7
8	9	10	11	12	13	14
15	16	17	18	19	20	21
22	23	24	25	26	27	28

FEBRUARY 2023

(The year of Sunyear)

Sunday	Monday	Tuesday	Wednesday	Thursday	Friday	Saturday
1	2	3	4	5	6	7
8	9	10	11	12	13	14
15	16	17	18	19	20	21
22	23	24	25	26	27	28

MARCH 2023

(The year of Sunyear)

Sunday	Monday	Tuesday	Wednesday	Thursday	Friday	Saturday
1	2	3	4	5	6	7
8	9	10	11	12	13	14
15	16	17	18	19	20	21
22	23	24	25	26	27	28

APRIL 2023

(The year of Sunyear)

Sunday	Monday	Tuesday	Wednesday	Thursday	Friday	Saturday
1	2	3	4	5	6	7
8	9	10	11	12	13	14
15	16	17	18	19	20	21
22	23	24	25	26	27	28

MAY 2023

(The year of Sunyear)

Sunday	Monday	Tuesday	Wednesday	Thursday	Friday	Saturday
1	2	3	4	5	6	7
8	9	10	11	12	13	14
15	16	17	18	19	20	21
22	23	24	25	26	27	28

JUNE 2023

(The year of Sunyear)

Sunday	Monday	Tuesday	Wednesday	Thursday	Friday	Saturday
1	2	3	4	5	6	7
8	9	10	11	12	13	14
15	16	17	18	19	20	21
22	23	24	25	26	27	28

A NEW ZEAR

JULY 2023

(The year of Sunyear)

Sunday	Monday	Tuesday	Wednesday	Thursday	Friday	Saturday
1	2	3	4	5	6	7
8	9	10	11	12	13	14
15	16	17	18	19	20	21
22	23	24	25	26	27	28

AUGUST 2023

(The year of Sunyear)

Sunday	Monday	Tuesday	Wednesday	Thursday	Friday	Saturday
1	2	3	4	5	6	7
8	9	10	11	12	13	14
15	16	17	18	19	20	21
22	23	24	25	26	27	28

SEPTEMBER 2023

(The year of Sunyear)

Sunday	Monday	Tuesday	Wednesday	Thursday	Friday	Saturday
1	2	3	4	5	6	7
8	9	10	11	12	13	14
15	16	17	18	19	20	21
22	23	24	25	26	27	28

THE NUMBER 28

OCTOBER 2023

(The year of Sunyear)

Sunday	Monday	Tuesday	Wednesday	Thursday	Friday	Saturday
1	2	3	4	5	6	7
8	9	10	11	12	13	14
15	16	17	18	19	20	21
22	23	24	25	26	27	28

NOVEMBER 2023

(The year of Sunyear)

Sunday	Monday	Tuesday	Wednesday	Thursday	Friday	Saturday
1	2	3	4	5	6	7
8	9	10	11	12	13	14
15	16	17	18	19	20	21
22	23	24	25	26	27	28

DECEMBER 2023

(The year of Sunyear)

Sunday	Monday	Tuesday	Wednesday	Thursday	Friday	Saturday
1	2	3	4	5	6	7
8	9	10	11	12	13	14
15	16	17	18	19	20	21
22	23	24	25	26	27	28

A NEW ZEAR

LOVEMBER 2023

(The year of Sunyear)

Sunday	Monday	Tuesday	Wednesday	Thursday	Friday	Saturday
1	2	3	4	5	6	7
8	9	10	11	12	13	14
15	16	17	18	19	20	21
22	23	24	25	26	27	28
29						

Notice how each week started on a Sunday, the year started on a Sunday, and Sunday is the last day of the year. Therefore, this is the year of Sunday. So, we can label this year as Sunyear; moreover, the following year will be label as Monyear. Obviously, the following year is Tuesyear, then Wednesyear, Thursyear, Friyear, Saturyear, and then start back over on Sunyear. Now you see how the years are forming a pattern that is exactly like the days of the week.

Notice how each month has 28 days in the year except one extra day at the end. This day is the key to link the following years together for the following reasons: if you know how to add, multiple, and do division, then notice how the earth is six hours away at the start of a new year. This is a total of 24 hours, which is a day in four years, which is why we have a leap year, right.

Now notice this same pattern in a year. Notice that ¼ of a year will also end in 13 weeks and six hours (that is, a frayation); moreover, since our calendar year and days start at midnight, then 13 weeks and six hours will start and end at midnight. The next ¼ of a year, which is ½ of a year, will start and end in the morning. The next ¼ of a year, which is ¾ of a year, will start and end in the afternoon. The next ¼ of a year, which is a whole year, will start and end in the evening; moreover, when the new year begin, these patterns, which is indicated with

THE NUMBER 28

a frayation, will start in its original position, which is midnight. There is also 52 weeks and a day (that is, four frayations) in a year. Now what I just mention is no new discovery, but it is something that you can see for yourself.

Notice the following facts with this new discovery regarding the correct number of days in a month, which is 28 days, and thirteen months in a year. This causes the days to form the same pattern as frayations in a year. The calendar, which was just previously illustrated, has 13 months and a day in a year, which is ¼ of a Tay. Since the first day of the year in the above calendar is on a Sunday, and the new year starts at midnight, then the last day of the year is on a Sunday; consequently, when the earth reaches its original position, it is 6 am on a Monday. The earth has shifted 90°. When the earth reaches its original position during the beginning of the third year, it is 12 pm on a Tuesday. The earth has shifted 180 degrees. When the earth reaches its original position during the beginning of the fourth year, it is 6 pm on a Wednesday, and the earth has shifted 270°. Now we are at the completion of four years, which is a Tay. The earth is a whole day from its original position; therefore, when the earth reaches its original position during the beginning of the fifth year, it is 12 am on a Friday, and the earth has shifted 360 degrees. Now let us make comparison between the pattern with the frayations and days: (1). There are 365 frayations in ¼ of year, and 365 days in ¼ of a Tay. (2). Both start and end on the same day (that is, in the same position of a day). (3). There are 28 frayations in a week. There are 28 days in a month. There are 13 weeks and a frayation in ¼ of year. There are also 13 months and a day in ¼ of a Tay. (4). Consequently, there are 52 weeks and four frayations in a year. There are also 52 months and four days in a Tay. (5). Lastly, there are 1,460 frayations in a year, and there are 1,460 days in a Tay (that is, four years).

I was considering another illustration of a calendar for the start of

A NEW ZEAR

the following year, which is the year, 2023. Then I notice that the last day of this year is on a Saturday. (I wonder if this is a coincidence, or a divine inspiration to put this book out now as the time to reveal this book to the world). So, I can use the previous calendar to serve two purposes, which is as follows: (1). Its serve as an example of the correct calendar year, and the calendar for 2023. The following calendar is for 2024. Notice the slight difference in the two calendars. Remember, this book will show how the years are forming the same pattern as the days and frayations.

We have Sunday, Monday, Tuesday, Wednesday, Thursday, Friday, and Saturday in a week, right. One evidence that the years are forming the same pattern as the days. Notice that the first and last day of 2022 is on a Saturday. This is supposed to be the year of Saturday (that is, Saturyear). Consequently, the first and last day of 2023 is on a Sunday, and the first and last day of 2024 is on a Monday, right (see how the years are forming a pattern as the days). This is the year of each week to start on a Sunday (that is, the year of Sunyear) in the above illustration. Every week of the month start on a Sunday, and the first and last day is on a Sunday. Again, the last day of the year links the years to keep this as a consistent pattern with the years. Notice if it was 364 days in a year. The year would always start and end on a Sunday; nevertheless, nature is designed with 365 days in a year. Therefore, since 2023 starts and ends on a Sunday, then 2024 will start on a Monday. The year with weeks to start on a Monday. This book labels the year of Monday as Monyear. One more point before I make an illustration of 2024. Notice that the first day of each month in 2023 in the first illustration started on a Sunday, and the first day of each month in 2024 in the following illustration started on a Monday, which is another evidence of a consistent pattern with this new discovery.

THE NUMBER 28

JANUARY 2024

(The year of Monyear)

Sunday	Monday	Tuesday	Wednesday	Thursday	Friday	Saturday
	1	2	3	4	5	6
7	8	9	10	11	12	13
14	15	16	17	18	19	20
21	22	23	24	25	26	27
28						

FEBRUARY 2024

(The year of Monyear)

Sunday	Monday	Tuesday	Wednesday	Thursday	Friday	Saturday
	1	2	3	4	5	6
7	8	9	10	11	12	13
14	15	16	17	18	19	20
21	22	23	24	25	26	27
28						

MARCH 2024

(The year of Monyear)

Sunday	Monday	Tuesday	Wednesday	Thursday	Friday	Saturday
	1	2	3	4	5	6
7	8	9	10	11	12	13
14	15	16	17	18	19	20
21	22	23	24	25	26	27
28						

A NEW ZEAR

APRIL 2024

(The year of Monyear)

Sunday	Monday	Tuesday	Wednesday	Thursday	Friday	Saturday
	1	2	3	4	5	6
7	8	9	10	11	12	13
14	15	16	17	18	19	20
21	22	23	24	25	26	27
28						

MAY 2024

(The year of Monyear)

Sunday	Monday	Tuesday	Wednesday	Thursday	Friday	Saturday
	1	2	3	4	5	6
7	8	9	10	11	12	13
14	15	16	17	18	19	20
21	22	23	24	25	26	27
28						

JUNE 2024

(The year of Monyear)

Sunday	Monday	Tuesday	Wednesday	Thursday	Friday	Saturday
	1	2	3	4	5	6
7	8	9	10	11	12	13
14	15	16	17	18	19	20
21	22	23	24	25	26	27
28						

THE NUMBER 28

JULY 2024

(The year of Monyear)

Sunday	Monday	Tuesday	Wednesday	Thursday	Friday	Saturday
	1	2	3	4	5	6
7	8	9	10	11	12	13
14	15	16	17	18	19	20
21	22	23	24	25	26	27
28						

AUGUST 2024

(The year of Monyear)

Sunday	Monday	Tuesday	Wednesday	Thursday	Friday	Saturday
	1	2	3	4	5	6
7	8	9	10	11	12	13
14	15	16	17	18	19	20
21	22	23	24	25	26	27
28						

SEPTEMBER 2024

(The year of Monyear)

Sunday	Monday	Tuesday	Wednesday	Thursday	Friday	Saturday
	1	2	3	4	5	6
7	8	9	10	11	12	13
14	15	16	17	18	19	20
21	22	23	24	25	26	27
28						

A NEW ZEAR

OCTOBER 2024

(The year of Monyear)

Sunday	Monday	Tuesday	Wednesday	Thursday	Friday	Saturday
	1	2	3	4	5	6
7	8	9	10	11	12	13
14	15	16	17	18	19	20
21	22	23	24	25	26	27
28						

NOVEMBER 2024

(The year of Monyear)

Sunday	Monday	Tuesday	Wednesday	Thursday	Friday	Saturday
	1	2	3	4	5	6
7	8	9	10	11	12	13
14	15	16	17	18	19	20
21	22	23	24	25	26	27
28						

DECEMBER 2024

(The year of Monyear)

Sunday	Monday	Tuesday	Wednesday	Thursday	Friday	Saturday
	1	2	3	4	5	6
7	8	9	10	11	12	13
14	15	16	17	18	19	20
21	22	23	24	25	26	27
28						

THE NUMBER 28

LOVEMBER 2024

(The year of Monyear)

Sunday	Monday	Tuesday	Wednesday	Thursday	Friday	Saturday
	1	2	3	4	5	6
7	8	9	10	11	12	13
14	15	16	17	18	19	20
21	22	23	24	25	26	27
28	29					

This is one sure evidence that a month is to consist of 28 days. Notice how many time that six hours can go into a week; it is 28 times; moreover, there is 13 weeks and six hours in ¼ of year, which, again, come out to be 52 weeks and a day in year. Notice how the first six hours and the last six hours in this 13 weeks and six hours is in the same position, e.g., if it is six O'clock am, and you were to add 13 weeks and six hours to this. Whatever day that you land on would be at six O'clock am. Remember how I brought to your attention with each year to start and end on the same day. Therefore, the previous evidence just revealed how six-hours, which is a frayation, and 24-hour hours, which is a day, are forming the same pattern by landing in same position.

    Six O'clock am represent the spring in a day. So, if you were to add another 13 weeks and six-hours to this season of spring, you would not believe what time of day it is. It is the exact same time of day. If we add another 13 weeks and six hours, the first and last six-hours of the second set six hours would start at 12pm and end at 12 pm, which represents summer, because 12 pm is the hottest part of the day. Now add another 13 weeks and six hours, and the first six hours will start at 6pm and the last six hours is at 6 pm, which is fall. Now add the last set of 13 weeks and six-hours, and it will start at 12am and end at 12

35

A NEW ZEAR

am, which is the coldest part of the day like winter.

If you study closely the days, months, years, six-hours, and other patterns that I have revealed in this book, you will notice that they correlate with one another to establish the same pattern, and each pattern are forming the four seasons. I just revealed how these six hours (that is, frayations) are forming the four seasons in a year, and how days are forming the four seasons within a Tay. I will demonstrate later how years are forming the four seasons by establishing four Deasons, which is a Zear.

You see, we are under the impression that the days are forming the four seasons in a year, but it is six-hours to form the four seasons of a day, and four seasons of a year. Now compare the pattern with these frayations in a year to the pattern with the days in four years, which is a Tay. Again, there are 28 frayations in a week. 13 weeks and a frayation in ¼ of a year, which indicate a season. 52 weeks and four frayations (that is, a day) in a year. There is also a total of 1,460 frayations in a year. Now notice how the days are forming the same pattern in four years with 28 days in a month. This would cause 13 months and a day in a year. This would come to a total of 52 months in four years. There are 1,460 days in four years. You will learn in this book that the years to form the same pattern, too.

THE DISCOVERY OF A TAY

MOREOVER, THIS BOOK will introduce new patterns of time (e.g., four years, which is label as a **Tay**[6]), which others have overlooked as they attempted to create a calendar to reflect the solar system, that is, the earth as it orbits the sun. Let go of your education of four years as a leap year and embrace four years as a Tay as I reeducate you to adapt a calendar to indicate the exact time, that is, reflect the exact position of the earth to orbit the sun. No correct solar calendar has ever existed in the history of humanity. Every person from Julian Caesar etc. have admitted the fact about their calendar system to not be exactly right; it is as close as they could understand. Everything that they say is correct. Their lack of awareness of a frayation (that is, the pattern with six-hours) is what threw off their calculation to be correct.

The earth would have rotated 360 degrees in four years, which is a Tay (that is, 1,460 days), before its completion. Remember, there are 1,460 positions around the sun, which are the beginning point for every day, week, month, year, Tay, Deason, Zear, and Comyetion. Four years, which is a Tay, symbolize the beginning of the day to start at each of these 1,460 positions. Consequently, the earth has rotated 360 degrees in a Tay to complete a cycle with the days; however, this pattern with the Tays must happen 365 time, which is 365 Tays (that is, 1,460 years). This is also known as a Zear. If we erase

A NEW ZEAR

any knowledge of a leap year from our memory, and just observe the pattern that nature is forming by allowing a new year to start at a different position, we will learn the following facts: a new year will start in 1,460 positions before it to start in its original position, which is a total of 1,460 years; consequently, 365 Tays would happen during this period, which is label as a Zear in this book. Remember, a day would have started at each 1,460 positions in four years, which is a Tay; however, the next Tay (that is, four years) symbolize a rotation for the earth to start a new day in each 1,460 positions, but there is one difference between the previous Tay and the next Tay. While the next Tay will start at a position, which is four frayations (that is, a day) after the previous Tay. Obviously, this is the second Tay. This pattern will happen 365 times, which is, 365 Tays (that is, a Zear); in other words, there are 1,460 frayations, that is, 1,460 positions for the start and end of each day as the earth to orbit the sun. A day will only start at only 365 positions of these 1,460 positions within a year; therefore, it takes four years, which is a Tay, for a day to start at each position. Moreover, it takes 365 Tays for a day to occur 365 times at each position, which is 1,460 years (that is, a Zear). There is a reason for this pattern, which occur every four years. Notice the two positions of the earth at the beginning and end of a year. If you just observe the pattern of the year without a leap year, you will learn the following fact: The position of the earth at the beginning of the year of the first year is the last position of the year at the end of a Zear, and the last position of the year of the first year is the first position at the end of a Zear. Nature is nourishing the earth that is almost like a pig on a spit. Nature begins its process of nourishing the earth at a different position at the beginning of the year. If you were to observe the last position of beginning this process, you would observe the beginning and ending of this process to kind of like reverse and flip.

Notice the similarity of 1,460 years has a similar pattern. Nature is taking care of each parts of the earth evenly during this process;

however, the process is not over at this point. We think that the process of nature to renew the earth is in a year with the four seasons. The process of nature is in more detail than that. It is even in more detail after the process of the Tays, which build up to be a Zear.

THE DAYS OF OUR LIVES

TWENTY-EIGHT DAYS IN a month, which this book will give evidence to prove this fact: in other words, days, years, and frayations are forming the same pattern, and weeks and months are forming the same pattern. Notice how there are 28 frayations in a week, thirteen weeks and a frayation (that is, 365 frayations) in ¼ of a year, and 52 weeks and four frayations, (that is, a whole day), which is, 1,460 frayations in a year. Again, days are forming the same pattern as frayations, e.g., if you erase your education about 12 months in a year, and each month to have different amounts of days in each month and observe the pattern that is created with 28 days in a month. This would add up to 365 days in a year as always; however, you would come up with 13 months and a day in a year, and 52 months and four days in four years. This is no coincidence that the weeks and months are forming the same pattern. Julian Caesar and others could not figure out the exact amount of days for each month, and they have all admitted to this mistake. We adapted their way with honest intention, because no one knew the correct way. One error with their approach is each month to have different amounts of days, e.g., some months have 30 days, other ones have 31, and one has 28 days. Notice how each minute has sixty seconds, each hour has sixty minutes, each day has 24 hours, each week has seven days, and there are 365 days in each year. The months should be consistent with time as well. I am not introducing 28 days in a calendar to make up a consistent pattern, I am revealing the actual consistent pattern in nature with time.

THROUGHOUT THE YEARS

OUR EDUCATION WITH time is not covering every aspect of time, because we are taught that time starts over to its beginning at the end of the year. Notice how this behavior is not true with other aspects of time, e.g., notice the first second on a clock. The next second will not start in the same position, right; moreover, so many seconds, which is sixty seconds are building up to become a minute, then a second will start at the beginning (that is, its first position). This is also true with minutes, hours, days, weeks, and months. For no day starts in the same position in the solar system as the earth orbit the sun. The beginning of the day will not start in its original position until after 1,460 days, right. If you did not observe the leap year, you would learn that nature is forming the same pattern with the years as the days; therefore, the following year is not the completion of the cycle of time; the years are adding up to form other patterns, too, which this book label as a Tay, Deason, and Zear. Just as the following day is not the complete cycle of time but one of the days in a year; the beginning of the following year is not the complete cycle of time but one of the years in a Zear. The fact that the years are not starting in the same position is proof that the years are not the same. Years are different from each other like seconds, minutes, hours, days, weeks and months. Since the years are forming the same pattern as the days. This book will encourage you to identify the years with a difference, for example, we may ask or say, "what is the day, month, hour, minute, or second (to mean what is the time)", because days are not the same; consequently, we should say, "what is the year"? We identify each day of the week with a different name until it begins its new period, which takes a week. Since there are seven days in a week, we have seven names for the days of the week. We must identify the years like the days, since the years are forming the same arrangement as the days. Nature teaches that the following year will not begin in the same position as the previous year. We know that this is a fact, which is why we invented the leap year. However, we should not create a system to conform nature to our limited understanding. We should not become complacent with our accomplishment but continue to strive to gain a full understanding until our understanding is in conformity with the laws of nature about time. A proper education does not stop at halfway of understanding anything; education stops at the full understanding of a subject. Then we must continue to educate ourselves about previous lessons, or we will forget it (that is, lose the knowledge that is gain).

This book will not look at a year to start over at the beginning of the cycle of time. It will accept a new year as the next year. This was the error with others who attempted to solve this mystery of time, which is why they introduced some type of leap year to adjust the earth to start the new year in the same position. The ideal that they came up with the leap year to accomplish this purpose, this is evidence within itself that the earth was not naturally producing this result. Therefore, I did not come up with a method to force nature to follow my way, I accepted this fact that the earth has shift from its original position from the previous year as a reality, and I just watch the pattern that it was forming. I was not trying to figure it out. My love and observation for life, and the value of time is how I discover the truth.

Let us move on! I want to discuss other familiar patterns of time to help you to understand the pattern with the years. Every second is just

A NEW ZEAR

a new second, and sixty seconds are a minute. In other words, a new second is not the beginning of the cycle of time (that is, the earth is not in its original position); these seconds are adding up to become a minute of time, because there are sixty seconds in a minute, right. The same is true with the minutes, that is, a new minute is also not the beginning of the cycle of time; these minutes add up to be an hour, because there are sixty minutes in an hour. Moreover, an hour is not the beginning of the cycle of time for the earth to be in its original position; hours add up to be a day, which is twenty-four hours in a day. The beginning of a day is not the completion of the cycle of time, because, again, the earth is not in its original position as the day before it. To make a long story short, these days turn into a week; weeks turn into a month; months turn into a year, right. This is the correct education; however, there is more to learn about time. We agree that a new second, minute, day, week, or month is not the beginning, because the earth is not in its original position, then we should accept the fact that a new year, too, is not the beginning of the cycle of time, because we know that it is a scientific fact that the earth is not in its original position at the end of a year (that is, the start of a new year). So, this question should rise in your mind ... "years are creating what"? Just like sixty seconds turn into minutes, sixty-minutes turn into an hour et cetera, I have discovered that years are turning into a Tay[7] and these Tays are developing a Zear. I laid this foundation to reveal that the pattern with the years is exactly as the days. There are seven days in a week: Monday, Tuesday, Wednesday, Thursday, Friday, Saturday, and Sunday, right. Seven is the sign of completion for the days, and eight is the new beginning of the days to start a new week, right. If the day is Monday, then eight days from now is on a Monday. The years are forming the same pattern, e.g., notice that the beginning and end of every year is on the same day. So, if the year starts on a Monday, then the last day of the year is also on a Monday, and the beginning of the new year is on a Tuesday. If you follow this pattern, it is obviously that seven years is the completion of one pattern with the years, because the eighth year starts on Monday, too, and there is 28 (that is, winter, spring, summer, and fall) seasons during this period. So, seven years is forming the same pattern as a week, and month, which is 28 seasons.

Remember, I show you how there are 28 frayations in a week, but what I did not show is how a frayation is like season. If you equally divided the day into four divisions, which is four frayations (that is, four divisions of six hours), you will notice that each frayation has the characteristics of the four seasons, because morning is like spring, the afternoon is like the summer, the evening time is like the fall, and midnight is the coldest part of the day, which is like winter. This is no coincidence but is relevant to time. But back to the years, and more on that later! We should not look at a new year as just a new year, but a new year that is different from the previous year, which is how we recognize the days. Since we identified the days by naming them with a different name. We need to identify the years in like manner to accurately understand, and interpret time correctly, e.g., since there are seven unique years, then we should name them as Monyear, Tuesyear, Wednesyear, Thursyear, Friyear, Saturyear, and Sunyear.

We just discussed the pattern with eight years, but let's back up to four years, which is label as a Tay. This book brought to your attention about the pattern with six hours, which are label as a frayation, e.g., we think that the four seasons in a year are determined or considered by the days of the year, right. However, it is determined by frayations, because there are 365 frayations in ¼ of a year. Therefore, the pattern of these frayations reflect and align with the four seasons of the year. Even how each ¼ of a year, which is 365 frayations, start and end in a day exactly as the four seasons. In other words, ¼ of a year, which has a total of 365 frayations, start and end in the morning; the next ¼ of a year will start and end in the afternoon; the next ¼ of the year will start and end in the evening, and the last ¼ of the year will start and end at midnight, which is a total of 1,460 frayations. I repeat, the days are not

A NEW ZEAR

forming this pattern with four seasons within a year, it is these frayations. The days are forming this same pattern as these frayations within four years, which is a Tay. For there are 365 days in a year (that is, a year indicate the earth to complete ¼ of its pattern with the days. Whatever is the position of the earth to face the sun at 12 O'clock on the first day of the year, when the earth reaches this position during the following year, the part of the earth, which is facing the sun, is six hours away (that is, a frayation). In other words, the earth has rotated 90° after 365 days, which is the first season. Now the part of the earth that was facing the sun during the previous year, which is the season of summer for the first year, is now in the position of fall during the following year. When the third year of a Tay starts, the earth has rotated 180°; in other words, if the location of the earth is in America to face the sun during the first year, then the location of the earth to face the sun is in China (whatever country is the opposite of America) during this year, and the time of day in America is midnight. Now when the earth begins its fourth year, the earth has rotated 270°; consequently, the earth would have rotated 360° at the end of the fourth year. Therefore, the new year will start the day in its original position of the earth to face the sun as the first year; however, the year will not start in its original position.

Therefore, the completion of a Tay (that is, four years) is not the completion of time. The completion of a Tay is the completion of the first degree of the pattern of a Zear[8]. More on the discuss of a Zear later in this book. Just like sixty seconds are adding up to be a minute; sixty minutes turn into an hour, twenty-four hours make a day, 365 days make a year, four years make a Tay, and 365 Tays (that is, 1,460 years) make a Zear. But one more point on the days! Notice how it is 365 days in a year, which is, ¼ of a Tay, and 1,460 days in a Tay. If you calculate the number of days in a year, and four years, you will discover that the days are forming the same pattern as the frayations and producing a similar result as the four seasons in a year: (1). 1,460 days and 1,460 frayation is the complete cycle, and the beginning of its cycle. (2). 365

days is a season in a Tay, and 365 frayations is a season in a year, and ¼ of the completion of its four seasons. Now I am about to reveal a new discovery about the years. I showed how the days and frayations are forming the same pattern with four seasons, and the first 365 days of a Tay, which is four years, and the first 365 frayations of a year indicate the first seasons; the second 365 days and second 365 frayations indicate the second season; the third 365 frayations and third 365 days indicate the third season; the last 365 frayations and the last 365 days indicate the last season, which is the completion of the four seasons.

Remember, let everything go that you have been taught about a leap year, and just observe what happen. If you observe the first and last day of a year on a calendar, you will learn that the day is the same. Consequently, if the first and last day of the year is on a Monday, then the following year is Tuesday, which is the year of Tuesday (that is, every week will start on Tuesday). The following year will start and end on Wednesday with each week to start on Wednesday. So, the years is forming a pattern like the days of a week, right. Now observe this behavior! You will not believe the position of the first 365 years, which is a deason[9]. Since a 360° angle is a full circle, the first 365 years will start and end in a season, which is a 90° angle of the earth to orbit the sun in a year (that is, ¼ of the area of the path of the earth to orbit the sun). Therefore, it takes 1,460 years, which is label as a Zear, before the start of a new year with the earth in its original position, and the original position of the earth to face the sun. What Julian Caesar and others was trying to discovery, it does not take place in four years, but 1,460 years.

I conclude by repeating myself and adding new information to that. I brought to your attention by observing the first and last day of a year on your calendar, which show that the first and last day of a year is on the same day for a reason. If you were to observe the pattern of the years without observing a leap year, you will notice that the years are forming a pattern and season like the frayations of a year. For there are 365 frayations in ¼ of a year, 365 days in a year, which is ¼ of a Tay,

A NEW ZEAR

and 365 years in a Deason, which is ¼ a Zear. ¼ of a year, ¼ of a Tay, and ¼ of a Deason indicate a season; therefore, it takes 1,460 frayations, which is a year for these frayations to complete its four seasons, 1,460 days, which is four years (that is, a Tay) for the days to complete its four seasons, and 1,460 years, which is a Zear, for the years to complete its four seasons. In other words, we have a leap year, because the earth starts a new year that is a day away from its original position, since the earth starts a new year that is six hours (that is, a frayation) away from its original position at the end of a year.

No one could understand this change at the end of the year, and at the end of four years; therefore, we adapted a leap year to cause the new year to start in its original position, which is every four years. Here is the reason why the earth is in a position that is six hours away at the start of a new year. If the earth is six hours (that is, a frayation) away from its original position at the start of a new year, the earth is starting a new year at each position of a frayation, which is 1,460 frayation (that is, 1,460 positions), e.g., if you were to take a calculation, and multiple one day by 24 hours, you will get 24 hours, right; consequently, there is 8,760 hours in a year. So, if the earth is six hours away from its original position at the start of the following year, then it is a day away from its original position in four years; moreover, if there are 8,760 hours in 365 days (that is, a year), and if you were to divide 8,760 hours by 6 hours, the total is 1,460. So, there is 1,460 frayations (that is, six hours) in a year, right. Consequently, Since the earth starts the following year, which is six hours (that is, a frayation) away from the previous year, then this is the evidence that there are 1,460 position to start each year before the earth to start a new year in its original position. If you were to follow the start and end of a day, week, and month, you will also observe that they also start and end at one of these positions. Therefore, this book label the positions between six hours as a frayation. These position are the foundation of the start and end of every day, week, month, year, Tay, Deason, and Zear.

THROUGHOUT THE YEARS

I strongly suggest that you follow the pattern of six hours to solve this mystery, because, again, if there are 8,760 hours in a year, you can divide 8,760 hours by six hours, which is 1,460. If there are 1,460 six hours (that is, 1,460 frayations) in a year, notice how each day begins and ends in one of these positions for frayation, each year begins and ends in one of these positions for frayation, and ¼ of a day begins and ends in one of these positions for frayation. Therefore, these 1,460 positions for frayations are major position for a whole, e.g., a clock has twelve positions for the hours (24 positions, that is, if you are using military time), and every hour always start in one of these twelve positions. These positions indicate a whole hour. These 1,460 positions of frayation indicate either ¼ of a day or whole day, ¼ of a season, or completion of all four seasons, and the start and end of a year. If the days, years, Tays, Deasons, Zears and any other pattern are always starting and ending in one of these positions, then we must observe these positions for its relevance to time to indicate them as key positions in time.

I notice something interested about time while studying it. There are literally 365 days in a year, that is, it might take 365 ¼ days to reach the original position, but only 365 days to reach all position around the sun. Since the first day of the next new year, which is 2023, is on a Sunday, I created a chart after this paragraph to show the position of the earth on Sunday, and the time of day when the earth will reach this position during the following years after 2023.

A NEW ZEAR

	Reach the original position
January 1, 2023, is on a Sunday	**Sunday at Midnight**
January 1, 2024, is on a Monday	Monday at 6 a.m. (6 hrs. away)
January 1, 2025, is on a Tuesday	Tuesday at 12 noon (12 hrs. away)
January 1, 2026, is on a Wednesday	Wednesday at 6pm (18 hrs. away)
January 1, 2027, is on a Thursday	**Friday at midnight (1/2/2027)**
January 1, 2028, is on a Friday	Saturday at 6am (1/2/2028)
January 1, 2029, is on a Saturday	Sunday at 12 noon (1/2/2029)
January 1, 2030, is on a Sunday	Monday at 6pm (1/2/2030)
January 1, 2031, is on a Monday	**Wednesday at midnight (2 days away)**
January 1, 2032, is on a Tuesday	Thursday at 6 a.m. on 1/3/2032
January 1, 2033, is on a Wednesday	Friday at 12 p.m. on 1/3/2033
January 1, 2034, is on a Thursday	Saturday at 6 p.m. on 1/3/2034
January 1, 2035, is on a Friday	**Monday at 12 at midnight (3 days away)**
January 1, 2036, is on a Saturday	Tuesday at 6 a.m. on 1/4/2036
January 1, 2037, is on a Sunday	Wednesday at 12 p.m. on 1/4/2037
January 1, 2038, is on a Monday	Thursday at 6 p.m. on 1/4/2038
January 1, 2039, is on a Tuesday	**Saturday at midnight (4 days away)**

If we would not apply a day to the calendar on each fourth year, and just watch how the years unfold, you will notice a very beautiful pattern above here. Therefore, you should study the pattern above here to understand the pattern that is formed with the years.

WHAT IS A DEASON

A DEASON IS label as 365 years in this book. You will comprehend the mystery about the end of a year and the beginning of the next year, which is six hours away from the beginning of the previous year.

Guess the position of the earth after 365 years without considering a leap year! The new year will start at ¼ of the distance away from the start of the first year. Notice this pattern with the years are identical with the pattern with the days and frayations. Just like the first and last day of 365 days of a year is on the same day, the first and last year of a deason, which is 365 years, is on the same day; moreover, let us say that the first day of the first year was on a Sunday, the earth will start the new year after 365 years on a Monday, then on a Tuesday at the start of the third deason, and on a Wednesday with the start of the fourth deason. These four deason is label as a Zear, which is discuss later in this book. But for now, I can inform you about the end of the book. There are 28 deasons in the complete process of time, which is a total of 10,220 years. I will give evidence about this fact, which is later in this book.

Everything with time is even: every minute has the same number of seconds, every hour has the same number of minutes, every week has 7 days, and every year has 365 days. This is a pattern with time. Nature also has established the seasons, months, and years with a consistent pattern, but our calendar does not reflect this, because some years have 365 days, while one will have 366. Some months have 31 days, some have 30 days, and one has 28 days. The reason why no one has been able to create a calendar to have the same amount of days in every month, the seasons to always start at the same time, and the years to always have the same amount of days, because no one has been able to comprehend a mystery about time, which is, why is the earth at a distant of six hours away at the end of the year? The mystery is this! Everyone's attempts to solve this mystery have viewed the beginning and end of a day as the main point (that is, position) of the earth to orbit the sun; however, one day is not just one point. One day has four major points in time. If you observe these four points/positions in a day, which are six hours a part, you will understand exactly why the earth is six hours away at the beginning of a new year. In other words, I have discovered that every six hours, which is a frayation, is a main point of time. Therefore, there are four main points in a day.

I discover every six hours as a main point for the earth to orbit the sun by observing the following facts about time: (1). The earth is six hours away from its original position at the beginning of the next year, which is a total of 24 hours after four years; therefore, we have the leap year, right. (2). I discover the truth by observing this behavior without observing the leap year. You will not believe this … the reason for the earth to not be in the same position at the end of the year, but six hours away, because it is forming the same pattern as the days and frayations. If you follow this behavior with the years to start a new year at six hours away from the previous year without observing a leap year, you will observe the years to form a pattern to reflect the four seasons, e.g., follow this pattern for 365 years. The earth would eventually start a new year, which is ¼ away from its original position. If you divide a year by four, which is ¼ of a year, you will get 365 frayations, and this position is also at the same position as the years after 365 years. 365 days, 365 frayations, and 365 years is ¼ of it cycle, and each ¼ represent the four

WHAT IS A DEASON

seasons, that is, to start and end during the four seasons.

Now back to these 365 years will occur during a season, which is label as a deason in this book. If you follow this patten until the earth to start a new year in the same position, this will happen at the end of 1,460 years, and the start of 1,461 years; in other words, there are 1,460 positions for the years to start a new year on the path of the earth to orbit the sun, and these positions are a frayation (that is, six hours a part). If you divide a year by six hours, you also get 1,460. I mention the next fact earlier in this book, which also go good here, too. Every day, month, season, week, year, and any other pattern that you find in time will start and end at one of these positional frayations, that is, I have found 1,460 positions, which every day, year, week, month et cetera will start, and these positions are six hours a part. Do not trust me but follow the following facts to see if this is true. There are 365 days in a year. Now calculate to see how many time that six hours will go into 365 days, which is 1,460. Now create 1,460 points/positions on a circle to represent six hours apart, and the circle to symbolize the path of the earth to orbit the sun. Now pick any day of the week, any week of the month, any month of the year, or any year; you will notice that all of them to always start at one of these positions of a frayation; moreover, nothing (for example, the day, season, week, month, or year) to start between a frayation (that is, a point of six hours). Therefore, six hours, which is label as a frayation in this book, is the foundation.

Now back to 365 years, which is a deason ... I ask you to keep the pattern in mind, which is designed by a frayation. This was discussed in the chapter about frayation. I will mention one fact about these frayations to show a comparison between them and 365 years. 365 frayations are ¼ of a year; moreover, the last and first frayation is in the same position of the day, e.g., since it is midnight when the year start, the last frayation is at midnight (midnight symbolize winter, which is the colder part of the day). The next ¼ of the year with these frayations will start and end in the morning (the morning has the characteristic

53

A NEW ZEAR

of spring). I do not have to mention the last two ¼ of the year, because you probably see where I am going with this. Now back to 365 years! If you were to observe the position of the first year, and the day of the week (e.g., let's say that it is on a Monday), you will learn that after 365 years that the year has started in a position of the earth to orbit the sun, which is ¼ away from the first year, and this year is on a Monday. This is another example of the days and years are forming the same pattern, because the last and first day of 365 days of a year is also on the same day. If these first 365 years were during the spring, you will not believe this, but the beginning of each year for these 365 years will occur only in the region for the season of spring. If it is spring, the next 365 years will start and end on the same day, and only occur in the area for the summer. The next 365 years will start and end on the same day, which is a Wednesday, and only occur in the area for the season of fall. The next 365 years will start and end on the same day, which is Thursday, and only occur in the area for the season of winter. This is a total of 1,460 years, which this book label as a Zear. The end of this year for 1,460 years, which is a Zear, will start the year in its original position as the first year; however, the day of the week is on Friday.

Please pay close attention to this … What others have tried to accomplish in four years to have the new year to start in its original position, this does not happen with nature until the completion of 1,460 years. However, even though nature is starting a new year in its original position, the day of the week for the beginning of this year is on a Friday. Remember, I said, "let's say that the first year is on a Monday. Therefore, the cycle of time is still not completed. When the earth starts a new year in its original position, and the first day of that year is on a Monday, I ask you to observe the pattern of time, which is a sure sign for the completion of the cycle of time, that is, the earth has completed its cycle of renewing itself. I will discuss this in the next chapter about a Zear. Remember, 365 years are a deason. There are 1,460 years, which are four deasons, before the earth to start a new year in its original position. Pay close attention to the number of deasons, when the earth to start a new year in its original position, and on the same day as the day of the first year, which is truly a sign of the completion of the cycle of time.

Since the first day of the year and the last day of a year is on the same day, that day determine what year we are in, e.g., if it is Monday, then the year is Monyear. Since years are forming the same pattern as the days, then it takes 365 years to form it's pattern, which is a season, and 1,460 years to complete all four seasons as it take 365 days to complete a season, 365 six hours, which is 1/4 of a year, to complete a season, and 1,460 days to complete its 4 season in 4 years, and 1,460 six hours (that is, frayation) to complete the four seasons in a year.

A ZEAR IS 1,460 YEARS

WHILE WE SEEK to create a calendar with the earth to start every new year in the same position, it will never happen, because a year is not the ultimate form of time. Just like the seconds, minutes, and hours in a day is not the final pattern. The day is the final pattern with the earth to rotate 360° in a day. Our education is not completed with time, because we think that the days, weeks, and months add up to the ultimate moment of time for the earth to make the same completion with a year. So, we try to create a calendar to reflect this inadequate knowledge about time. It is really the days, weeks, months, years, Tays, and Deasons together to complete this pattern, which is a Zear. The earth has started a new year at every position, and it is now starting a new year at its original position at the end of a Zear, which is 1,460 years. Therefore, we need a calendar to incorporate a Zear, because at the beginning of a Zear, the earth reflects time to always starts at the beginning of the second, minute, hour, day, week, month, year, Tay, and Deason, and it only happen at this point of time. The only thing that is not at its original point, and that is what day it is. A Zear is the ultimate form (that is, pattern), which indicates the earth to rotates 360° to start a new year in its original position.

Now if there are 1,460 intervals of six hours in a year, and the earth shifts six hours at the beginning of each year from the prior year.

A ZEAR IS 1,460 YEARS

Notice the pattern with the years. The earth will also shift 1,460 times before a new year to start in its original position.

Even though the earth is starting a new year in its original position at the completion of a Zear, everything else is not at its original position, e.g., if it was 3:30 p.m., we are at the beginning and original position of a new second, and minutes, but not with the hour and day. Likewise, 1,460 years is when the earth to start a new year in its original position, and the minutes, hours, days, weeks, and months, are at the beginning, but the day is not the same. If the first year was on a Sunday, the day is Thursday during the start of a new Zear. Pay close attention in this book when the earth to start a new year in its original position, and the indication of the pattern to determine the complete process of time. Remember earlier in this book, I showed how the years are forming a pattern as the days, e.g., it takes the completion of seven days before the same appear again. This is also true with the years, that is, it takes the completion of seven years before a year to start a new year on the same day as the first year. These Zears are forming the same pattern, but in more detail. It takes the completion of seven Zears before the start of a new year is on the day as the first year. However, what is more interesting about when it occurs with the Zears is this: Even though the first day and eighth day is on the same day; however, the location of the earth to start the eighth day is not in the same position as the first day. The same is true with the years. The first and eighth year is on the same day, but the location of the start of the eighth year is not in the same position as the first year; however, the first and beginning of the eighth Zear is on the same day as the first and beginning of the first Zear; therefore, the first and eighth Zear is in the same position, which is the beginning of time.

I notice something else significant about six hours (that is, a frayation). We are aware that the earth is six hours away from its original positions at the start of a new year, right. If you were to follow this pattern for four years, which this book label as a Tay, you will see these

57

A NEW ZEAR

Tays to form a pattern that is exactly as the days. For there are 365 days in a year, right. I ask you to follow the pattern of the year to start a new year, which is six hours away from the previous. If the earth is six hours away at the end of one year, then it is a whole day away at the end of four years, which is why we have a leap year; however, I ask you to not observe the leap year, and just watch the pattern that is form with the years. Since the earth is a whole day away after four years, which this book label as a Tay, guess how many time for the occurrence of this pattern with Tays (that is, four years) to take place before a new year to start in its original position? You are not going to believe this! The answer is 365. Do not take my words for it, you can follow the pattern with the Tays (that is, four years) for yourself! Since one Tay is four years, then 365 Tays are 1,460 years, which this book label as a Zear[10].

So, when the year of 1,461 begin, the second, minute, hour, day, year, Tay, deason, and Zear is line up at the beginning of its cycle in time; however, the earth has not quite completed its cycle. If the original day was on a Monday, when the year of 1,461 begin, the first day of the year at the beginning of next Zear is on a Friday. Before I go any further, let's back up a little. Each year begins and ends on the same day, right; therefore, if the year, which is composed of 365 days, started on a Monday, then the following year will start on a Tuesday. If you follow the pattern of the years, the first year, and 365 years later will start on a Monday, and the following year will also start on a Tuesday. Consequently, the fourth year will start on a Thursday, and the fourth set of 365 years will also start on a Thursday. This is how I was able to determine the exact day for the beginning of the year, which is 1,461 years from now, is on a Friday. This is evidence that it might be a new second, minute, hour, day, year, Tay, Deason, and Zear, but not the same day as the first day of the first year; so, this is not the beginning of a new cycle with time. When the earth eventually starts a new year in this position, and the year starts on a Monday, it would have completed the same pattern as a week, and a month. This is another proof

A ZEAR IS 1,460 YEARS

that the days and years are forming the same pattern. Follow me alone, and I will show you! Remember, we are at year 1,461, and the start of this year is on a Friday since the end of the Zear ended on a Thursday.

Now if you were to follow this pattern for another Zear (that is, 1,460, years), the year will start on a Tuesday. You do not have to go through each year to figure this out. That would take up too much of your time. I did that for you. Allow me to show a shortcut to you.

If the day is Monday, when the earth is in its original position, and a new year will shift to start the next year on the next day, which is a Tuesday, and so forth with this pattern. Since the years and days are forming the same pattern, and every seven days and seven years is like a week to start the next day and year on the same day as the first day. Consequently, if 1,460 days indicate the day to start in its original position; however, if the first day of the year was on a Monday, the beginning of the fifth years (that is, day 1,461) is on a Friday, then the beginning of the fifth Deasons (that is, 1,461 year) is also on a Friday. This is how you can easily follow the pattern of a Zear, since these pattern resemblance other patterns, or you can do it the hard way by adding these number up to see if I am correct for yourself, which is what you will have to do to really know if I am correct instead of taking my words for it. Now follow this pattern with a Zear for another seven time. The new year, new day, new minute, new second, new year, new Tay, new Deason, and a new Zear to start in its original position, and original day, which is a Monday;

While you are reading after this sentence, follow along by observing the following drawing. The star in the middle represents the sun. The inner-most circle, which is next to the sun represents the direction of the earth to move counterclockwise to orbit the sun. For we are taught that the sun raises in the east and set in the west; however, the sun does not move. It is the earth that is moving around the sun in a counterclockwise motion. Notice that I have the arrows pointing in the opposite direction for the second circle. When the year has finished

A NEW ZEAR

after 365 days, the earth is six hours away from its original position. The earth rotates six hours away at the start of a new year; however, this pattern is formed in the opposite direction then the pattern of the days of a year. The days are forming a pattern that is counterclockwise, while the years are forming a pattern that is clockwise.

It takes 1,460 frayations for the frayations to complete its purpose in a year, which are the four seasons in a year. It takes 1,460 days for the days to complete its purpose in four years, which is another type of four season with the same characteristic as the four seasons in a year. It takes 1,460 years for the years to complete its purpose, which is four deasons (four deasons also has the same characteristic as the four seasons in a year). Again, 1,460 years is a Zear.

One reason why Julian Caesar could not discover a calendar to reflect the exact movement of the earth to orbit the sun, because they did not think or know that there were more to time after a year. Another way to say it is, Rome once thought that the earth was flat, and it ended at a certain distant in the ocean from Rome; however, they found out that there was more to the earth than their view from Rome. As a matter of fact, they found out that the earth was round. Yet studying, researching et cetera help them to learn the truth. Well, I have study, did research, observe et cetera, and I have learned that there is more to time than a year. Notice how everyday does not start or end in the same position on the path of the earth to orbit the sun. Since the earth is 90 degrees from its original position after 365 days, then the beginning of a day will not start in its original position unit the end of 1,460 days. Guess what ... This is also true with the years. If we stop tampering with nature by creating a leap year to force nature to fix to our understanding, and let nature takes its course, we will learn the following truth. Again, basic knowledge of the solar system reveal that no day start and end in the same position of the earth to orbit the sun. It is also a known fact that the beginning of the next year does not start in the same position as the previous year.

A ZEAR IS 1,460 YEARS

If you were to follow the pattern of the earth to start a new year, which is six hours away from the previous year, you will learn that it also takes 1,460 years before a new year to start in its original position. This pattern with the years is the same as the days, because, again, a new day will not start in its original position until the completion of 1,460 days, which is four years, and the first day of the year is on a Thursday, that is, if the first day of the first year was on a Monday. Moreover, you can get 1,460 six hours (that is, frayations) out of 365 days. However, six hours does not occur (that is, start and end) in the same position of the course of the earth to orbit the sun. This is just more evidence that the beginning of a new year is not supposed to start in the same position, but it is starting a new year in the next position. The years will continue this pattern with the earth to orbit the sun until the years have completed it cycle to produce the four seasons, which is 1,460 years.

Notice after every 60 seconds, the second in time is line up at its beginning position, but the hour and minute hand are not at its first beginning. The minutes is at the next position, and the hour hand has gradual moved to its next position. When 60 minutes have occurred, then the minute and second are at its original position, but the hour hand is still not at its beginning position, but at its next position, right. For the hour to begin at its beginning position, 24 hours must pass for this occurrence; however, we are not at the beginning of the week, month, and year. By now, I hope that you get my point, because I cannot stress enough to no longer view a new year as the end of the cycle of time, that is, the seconds, minutes, hours, days, weeks, months, frayations, and years have completed 360° to be back at it beginning, but the patterns, which are established with and/or after years, are not at their beginning position, which are a Tay, deason, Zear et cetera. The days are creating a Tay, the years are creating deasons, and Tays are creating a Zear. The end of four years, which is 1460 days, is the beginning of a new second, minute, hour, day, week, month, year, and Tay, but a

A NEW ZEAR

Tay, which is four years, is not the end of the cycle of time. For Tays are forming a Zear, and the years have not completed its cycle, which establish deasons. It takes 365 Tays, which is 1,460 years, for Tays to complete a pattern, which is a Zear, and 365 years for the years to complete ¼ of its pattern, which is a deason. 1,460 years is when the seconds, minutes, hours, days, weeks, months, years, Tays, and Deasons, have completed their cycle, which is a Zear; however, this is still not the complete cycle.

Another sign that time has not completed its cycle is a new year has started in the same position as the first year, but not on the same day of the week as the first year. The reason why I am pointed this out, because you are going to be amaze at the point when it occurs, and you are going to be amaze about the design of time. It is so beautiful, and perfect. Please be patient ... I am just as excited to tell you, then you are to finally know the mystery of time.

THE COMYETION [11]

I BROUGHT TO your attention about observing the first and last day of a year, which is on the same day. Let's say that this day is a Monday to draw my conclusion. If you observe 365 years, the first day and last day is on a Monday, and these 365 years will occur in the location of ¼ of the path of the earth to orbit the sun. You can conclude that the next 365 years will occur on a Tuesday, which indicates 180° of the years to complete its cycle. The next 365 years will occur on a Wednesday, which indicates 270° for the years to complete its cycle. The last 365 years will occur on a Thursday, which indicates 360° for the years to complete its cycle. Therefore, it takes 1,460 years for the years to complete its cycle, and to finally start a new year in its original position, which is called a Zear. I label these 365 years as a deason, which rhythm with season, because these four deasons occur in a specific area of the path of the earth to orbit the sun, and each deason starts and ends in the region of a season. Since the last 365 years start and end on a Thursday, then the day is on a Friday as the earth starts a new year in its original position. I hope that you understand the pattern of 365 years to create a deason, and four deasons create a Zear, which indicate the years to complete its cycle to begin a new year in its original position.

You will be wise to not just consider this moment as nothing, because there is a reason for the years to function in this way. Keep in mind that nature is keeping the earth in balance, and this is the process to evenly nourish the whole earth. Do not just look at time as nothing, but for its purpose. Nature is taking care of the earth. Time indicates this process of nature to nourish the earth, e.g., we know that the earth is six hours away from its original position (that is, at the beginning of a new year, right); however, I ask you to pay attention to this event. This event causes the earth to change position. For when the earth reaches its original position during the following year, the position of the earth is in a 90° different angle, when compared to the previous year; therefore, this part of the earth is now facing the sun in this location as the earth begin to orbit the sun during the following year. It takes four years for the whole earth to experiencing this process; however, this is just a process with the days. We have four seasons in a year, which indicate the process of the frayations. The four years, which we so-called a "leap year", indicate the days to process through the four seasons. 1,460 years indicate the years to process through the four seasons, which are four deasons, and a Zear. Four Zears, which is 1,460 Tays, indicates the Tays to process through the four seasons.

In other words, nature is caring for the earth via the years in the same manner, but in more detail than the days, which is why the earth is not in the same position at the end of the year. It takes a longer process than a year for nature to care for the entire earth. Nevertheless, we are almost at the end of this process. I ask you to keep this treatment of earth via nature in mind as we approach the end of the process of time, which indicate the earth to receive a full treatment of nature to nurture it … I was just mentioning the completion of a Zear. Let's pick back up from there! Notice how a day can be divided into four parts, which has the characteristic as the four seasons. There are four seasons in a year. Seven days fulfill the requirement of a week. Remember, I showed how these four frayations in a day to have the characteristic of a season; therefore, there are 28 of these patterns in a week. I also reveal how 28 days in a month is a consistent pattern in time with the weeks, and seasons. For there are 28 seasons in seven years. This is no coincidence, but it has the same purpose as the days. For there are seven days in a week; therefore, the completion of seven days is obviously the beginning of the eighth day, and that day is obviously on the same day. The same is true with years. The first and eighth year start the year on the same day; moreover, these seven years has 28 seasons as seven days has 28 unique pattern, which resemblance the four seasons. 28 days will also form the same pattern, because it is composed of four weeks, and it also has seven days in a week to form a pattern that resemblance a week, and seven years.

If you observe only deasons, Tays and Zears from here on out, you will notice how these Deasons, Tays and Zears together are forming the same design as a week, month, and seven years. If 1,460 years is a Zear, and the first day of the year during the start of 1,461 years later is on a Friday, which is when the earth to start a new year in its original position, let's just follow this pattern to observe its conclusion. Remember that one Zear has four deasons. Now this is the beginning of the second Zear (that is, second set of 1,460 years). The third set of Zear to start a new year in its original position is on a Tuesday. The fourth set of Zear will start the year in its original position on a Saturday. The fifth set of Zear will start the year in its original position on a Wednesday. The sixth set of Zear will start the year in its original position on a Sunday. If you study this pattern closely, you will observe the occurrence of 28 deasons. You will also observe that seven Zears have occurred, too; moreover, observe the chart that I have created to demonstrate this pattern. You will also notice that the days to never repeat itself in the same region of a deason, but all seven days. So far, we are at 10,220 years. Now we are to begin the eighth Zear, and the day is on a Monday. This is a very important moment in time, because never during the process of time has a new year to start in this position on a Monday except the first year. Please, do not look at time as nothing, but for its purpose.

A NEW ZEAR

If you observe this whole process, this is the only time for the earth to fully rotate 360° to each position around the sun. The only time when the earth to start a new year in the same position on the same day as the first year of this process. It occurs at the same time when the earth has move to each position and rotated 360 degrees to each position; in other words, if you do the correct calculation, this is the time when the whole earth has been process through nature to care for mother earth. Wow!!!!!!!

In other words, it takes seven Zears, which is a total of 10,220 years before the earth to start a new year in the same position on the same day as the first year of these 10,220. Therefore, again, the complete cycle of time is in seven Zears, which is a Comyetion, instead of a year. Notice that each Zear is forming a pattern like the days, years, and frayations. Remember, I demonstrated how there are 28 frayations in a week, 28 days in a month; I never explained how 28 years are forming the same pattern. That is obviously, since the years are forming a pattern like the days and frayations; however, I will take the time to explain a fact about 365 years, which is label as a deason. There are 28 deasons in the completion of the cycle of time, that is, when the earth starts a new year in its original position, and it will start a new year on the same day as the first year. Everything that I have discuss thus far can be supported with evidence beyond a doubt; however, my next point cannot be back with enough evidence. Even though everything is aligned at this point of time, I really believe that this is not the end, because it ends with 28 deasons (that is, seven Zears). If you continue after this point, everything seems to repeat itself in time, which is a strong indication that this is the complete cycle of nature. However, since it ended with seven Zears, is there really 365 Zears, or 1,460 Zears to complete the cycle of time. Again, I do not have enough evidence to prove this. Yet I have evidences, which is just mentioned to suggest this. However, my calculation up to seven Zears is adequate. I notice nothing to develop or change into something else after this point of time (the movement of the earth begins to repeat itself at this point, which indicates the end). I am still curious with how it is ending. I could be wrong, but the reason for my suggestion that there might be more to learn, because seven suggest a completion with other patterns; however, it always takes more than one completion of a pattern with seven to complete another pattern.

Remember earlier, I showed you how a day is like the four seasons, and there are seven days in a week, which is like 28 seasons with the frayations in a week, and there are 28 days in four weeks. I conclude with since these seven Zears have a patten like a week, and time starts in its original position after seven Zears, and the year starts on the same exact day of the week as the first year (no other point in time does this event takes place). There is something significance about this moment after observing the pattern that is formed with time at this moment in time. Yet I leave you with the next mystery of time. Does time start anew in seven Zears in our solar system and the entire galaxy (that is, all creation).

When you barbecue meat, you turn the meat to make sure that it cooks evenly on all sides and the inside. This is the pattern of the earth with its turning and being in a different position at the end of the year. Just imagine if the earth set still. One part of the earth would over cook from the sun, and the other part would probably freeze. Even with the earth not just turning, but also in a different position yearly. Its turning daily and being in a different position when it starts a new year is nature way of equally nourishing each part of the earth, and to keep the earth in balance. Look at the earth with your eyes. It's balance, right. I cannot reiterate enough that we must realize and accept the facts about the years are not functioning as we thought. Every calendar created by humanity (even Julian Caesar, which is the closely to the truth) are attempting to come up with a method for the years to always start in the same position in the solar system; however, nothing with time works like this. The seconds, minutes, hours, days, and months do not start

A NEW ZEAR

the next one in the same place but change position to indicate the earth to change position on its course to orbit the sun. Each part revolves in a 360 degrees circle before it starts in the same position. If we were to study the behavior of the years, we will notice that it naturally start a new year in a new position to revolve around the sun. So, we must end these leaps years to indicate that the earth is starting its new year in the same position (which it is not), and create a calendar to reflect the pattern of the movement of the earth to naturally revolve around the sun.

The earth is balancing itself with time. Notice, again, if the earth was to set still with America to face the sun. China (along with other countries in that region) will freeze for lack of sun. America would burn hotter than the desert for constantly facing the sun. Even with how today's theory of the earth, which is wrong, will cause the earth to be off balance, that is, if it was accurate. Nature is keeping the earth in balance, e.g., whatever is the position of the earth at the beginning of a new year to face the sun, notice that when the earth reach this position during the following year, it is now six hours away from the previous position (e.g., whatever is the position of the earth at 12'o clock a.m. in New York on New Year's Eve (the time in Hawaii will be 6'o clock p.m.). However, when the earth reaches this position during the following year, it will be the beginning of a new year in Hawaii, and the time is 6 a.m. in New York.

FIVE POINT STAR

THERE ARE 365 days in a year. Did you know that 365 is almost a prime number? The factors of 365 are 1, 5, 73, and 365. There is something unique about the factors of 365, because they create five points around the sun as the earth orbit the sun; in other words, 365 days divided by 5 = 73 (so, 73 days are 1/5 of a year), and 360° divided by 5 also equal 72° (so, 72° is 1/5 of 360°). Now I will show how 1/5 of 360° and 1/5 of 365 days correlate to time. We have heard some mystical story about the five-point star. Somehow the truth gets lost, and we begin to believe more in the fable about the truth, than the truth, itself. For example, we know the truth about Christ Jesus. Yet the Christmas Story about Santa Claw, the Easter Bunny and other fables about the birth and resurrection of Christ Jesus tend to overshadow the truth about CHRIST JESUS. The same is true about the five-point star. We have heard so many stories and meaning behind the five-point star, that the truth about the five-point star become lost. Nevertheless, the truth about five-point star relates to time. Each 73 days indicate 1/5 (that is, 72°) of the earth to orbit the sun. 73 days not only indicate 1/5 of the earth to orbit the sun, but what happen at this point of time, and where and when it takes place during the day.

I initially decided to not mention this about the five-point star, because I neither understood its arrangement, nor what is being designed

A NEW ZEAR

of one day and twenty minutes to indicate one degree of the movement of the earth to orbit the sun. It started over twelve years ago or so, when I begin to figure out most of time. However, here it is March 28, 2022, which is when I grasped the full understanding of time, that is, the earth in the solar system (well almost everything). I called my publisher to let them know that I will submit this book on my birthday, when I will be 52 years old. Since 52 is an important number in time, I interpreted my birthday as the time to submit this book; however, my birthday was not the time to submit this book, but when the full revelation of time came to me. Allow me to deviate from the purpose of this book for a moment to share something with you. I was homeless on my birthday with no one to share this day with me. I begin to read my Bible, and to thank GOD for another year. I was just thinking about life. I drifted into a deep sleep, then I begin to see vision of the great pyramid of Egypt. The sun appears to raise and fall over this pyramid. When I awoke, this dream prompt me to calculate the time and degree of the earth in a day. This made me to become curious about the mysterious point of time, which is the end of the year. No one has been able to explain it. I was not trying to figure out this mystery. I was just enjoying observing life, and time in relationship to it. New information would reveal itself to me about time on my following birthdays.

Last year was the first time that I did not spend time alone on my birthday. I decided to go out to celebrate my birthday. Nothing was revealed to me about time. I decided to spend time with just GOD on my birthday for this year to rededicate my life to GOD, be thankful for another year et cetera. This is an old ritual that I found in the first chapter of the book of Job. My point in sharing this with you is this … I think that there is something more to our birthday, because when I isolate myself on birthday to meditate on the significant of my birthday, which indicate the earth is exactly in the same position as it was on my birthday, I begin to have these awesome visions about the universe. I hope that this inspire you to try this ancient ritual on your birthday, because I want to know if this is a coincidence, or will this happen with anyone who try this ancient ritual.

So much knowledge is lost, which is why I entitle this book as "Lost Time", e.g., I thank GOD for the Torah, because I had a gut feeling that there are some lost ancient holy writings. The Torah help me to prove that my intuition was right. The book of Job was written 5,000 years before any other books in the Torah, and the author of the book is not a Jew. Why did the Jews put this book in their Torah, which is the only book in the Torah with an author who is not a Jew? I do not know, but I am glad that they did. This book proves that there was holy writing in existence before the existence of Israel. However, this could not be the only book. Where are the rest of these books? The only writing that we have are from the Jews, which books I love, and study dearly. However, Melchizedek was alive during this time. The Torah and Bible agree that Melchizedek was of the highest order of priest. So, if there is a higher order, and there are writing about this order, where are these books. For there are lost ritual that we do no practice, which may open up our spiritual understanding, open up our heart and mind about life, and enlighten us more about life, e.g., it was by practicing this ancient ritual in the book of Job, which reveal how they use to celebrate birthdays, how they would focus their attention on their birthday, how they would purify their soul on their birthday, and the enlightenment that was revealed on their birthday.

How did the Egypt create the great pyramid, how did they discover the hourglass, how the people from the region of the book of Job was able to understand the star to follow the stars to Jesus, and how they use the stars to interpret the time? For we do not know how to interpret the time. Why is the weather changing? I end with this point! If CHRIST JESUS assumed the order of Melchizedek, and the Torah record an event with Melchizedek to blessed Abraham, this is proof of a higher order of life. There is no book in the public bookstore about this order, and there is no book in the public during the time of Melchizedek

A NEW ZEAR

except the book of Job. There are some secrets that are going on about life around here! Either these books are lost or hidden from the public. Thank you for allowing me to share this personal point, since it is not relevant to the point of this book. However, I am on a journey for the truth, and I ask you to join me to help each other to learn more about life and time. Now back to the point of this book!

(Another mysterious occurrence, which I did not understand, came to me on my birthday!)

Since there is 24 hours in a day, and 365 days in a year, then there is 8,760 hours in a year (that is, 525,600 minutes in a year). Another way to say it is there are 365 days, which is a total of 1460 frayations. If there are 360 minutes in six hours (that is, one frayation), then you get 360 minutes in a frayation; therefore, one minute of a frayation is one degree. I mention this before now, because I have learned that six hours, days, hours, weeks, months, and years form the four seasons, and a position within the four seasons. Moreover, I just learned that even the minutes are playing an important role to reveal certain point (that is, position) of the earth in the solar system. This chapter will discuss the movement of the earth. The earth turns 360% in a day, and the earth orbit the sun in 365 days; therefore, if it takes 365 days for the earth to make a full circle of 360° to orbit the sun, then it takes one day and twenty minutes, which is 1,460 minutes, to complete 1° of its orbits around the sun. If there is 1,460 minutes in one degree of the earth to orbit the sun, then there is 365 minutes in ¼ of a degree of the movement of the earth to orbit the sun. Notice how even the minutes are forming the same patterns as the frayations, days, years et cetera, because there is a total of 365 frayation in ¼ of a year, and 1,460 frayations in a year. There is also 365 days in a year, and 1,460 days in four years, which is a Tay; moreover, there is 365 years in a deason, which was discuss earlier, and 1,460 years in a Zear, which was also discussed earlier. But back to the pattern with minutes!

This is how I arrived at my conclusion! You can divide 525,600 minutes, which is the total number of minutes in a year, by 360 degrees, which is 1,460 minutes. So, every 1,460 minutes, which is one day and twenty minutes, the earth move one degree to orbit the sun.

Therefore, 365 minutes (that is, six hours and five minutes) is ¼ of a degree of the earth to orbit the sun. Notice how the number 365 is the result of a pattern with the minutes, frayations, which is six hours, days, years et cetera to reflect ¼ of a pattern, that is, a season, and 1,460 is the result of a whole degree, that is all four seasons. Each falling into synchronization at the number 365 and 1,460. If it takes 1,460 minutes for the earth to slide to a new degree, and 1,460 minutes are a total of one day and twenty minutes, notice the point of time when the earth reach a degree at a whole day in the following chart; moreover, notice the position in comparison of the 360° of a frayation. Since the earth orbit the sun in 365 days, the first column represents the amount of time for the earth to move to orbit the earth. The second column represents the degree of the earth in the amount of time of the first column. The third column represents the position of the earth in relation to a frayation.

Consequently, every 73 minutes represent .05° of 360° of the earth to orbit the sun, and represent each degree (that is, each minute) in a frayation, which is, six hours. Moreover, every 365 minutes represent five-point star in relationship to a frayation, because every 73 minutes, the earth move five degrees in relation to a frayation. The diagram below reveals the following fact, too. If it takes one day and twenty minutes for the earth to move 1° on its course to orbit the sun, notice that the earth never finish or begin a degree at the beginning of a day except every 73 days. Moreover, 73 days is the only number of days that can be divided into 365 days, which is five time (not including 365 days and one day of course). This is no coincidence, because these 73 days design several patterns, and it also indicates the completion of these patterns. I just mention one pattern, which is 73 minutes. Now observe each 73 days! Notice how every 18 days and six hours, which

A NEW ZEAR

is 73 frayations, the earth is at the end position of the 18th frayations of the 1,460th frayations. Notice that this pattern occurs four times within 73 days. You should notice by now that every discovery of a pattern of time has four divisions, which indicate the four seasons. Therefore, every 73 days is significant to indicate 1/5 of the movement of the earth to orbit the sun, and completion of a pattern, which indicate four seasons. Since 365 days divided by 73 days is 5, then there are five of these position on the course for the earth to orbit the sun, which is why I label these points as five-point star.

THE EARTH ORBIT SUN	IN DEGREES	FRAYATION POSITION
73 minutes (that is, 1 hr. and 13 min)	.05°	—
146 minutes (that is, 2 hrs. and 26 min)	A tenth of a degree	—
219 minutes (that is, 3 hrs. and 39 min)	.15°	—
292 minutes (that is, 4 hrs. and 52 min)	.20°	—
365 minutes (that is, 6 hrs. and 5 min)	¼ °	5°
730 minutes (that is, 12 hrs. and 10 min)	½ °	10°
1,095 min. (that is, 18 hrs. and 15 min)	¾ °	15°
One day and twenty minutes	1°	20°
Two days and forty minutes	2°	40°
Three days and one hour	3°	60°
4 days and 80 minutes	4°	80°
5 days and 100 minutes	5°	100°
Six days and two hours	6°	120°
7 days and 140 minutes	7°	140°
8 days and 160 minutes	8°	160°
Nine days and three hours	9°	180°
10 days and 200 minutes	10°	200°
11 days and 220 minutes	11°	220°
Twelve days and four hours	12°	240°
13 days and 260 minutes	13°	260°

FIVE POINT STAR

14 days and 280 minutes	14°	280°
Fifteen days and five hours	15°	300°
16 days and 320 minutes	16°	320°
17 days and 340 minutes	17°	340°
Eighteen days and six hours	18°	**360°**
19 days and 380 minutes	19°	20°
20 days and 400 minutes	20°	40°
Twenty-one days and seven hours	21°	60°
22 days and 440 minutes	22°	80°
23 days and 460 minutes	23°	100°
Twenty-four days and eight hours	24°	120°
25 days and 500 minutes	25°	140°
26 days and 520 minutes	26°	160°
Twenty-seven days and nine hours	27°	180°
28 days and 560 minutes	28°	200°
29 days and 580 minutes	29°	220°
Thirty days and ten hours	30°	240°
31 days and 620 minutes	31°	260°
32 days and 640 minutes	32°	280°
Thirty-three days and eleven hours	33°	300°
34 days and 680 minutes	34°	320°
35 days and 700 minutes	35°	340°
Thirty-six days and twelve hours	36°	**360°**
37 days and 740 minutes	37°	20°
38 days and 760 minutes	38°	40°
Thirty-nine days and thirteen hours	39°	60°
40 days and 800 minutes	40°	80°
41 days and 820 minutes	41°	100°
Forty-two days and fourteen hours	42°	120°
43 days and 860 minutes	43°	140°
44 days and 880 minutes	44°	160°
Forty-five days and fifteen hours	45°	180°

A NEW ZEAR

46 days and 920 minutes	46°	200°
47 days and 940 minutes	47°	220°
Forty-eight days and sixteen hours	48°	240°
49 days and 980 minutes	49°	260°
50 days and 1,000 minutes	50°	280°
Fifty-one days and seventeen hours	51°	300°
52 days and 1,040 minutes	52°	320°
53 days and 1,060 minutes	53°	340°
Fifty-four days and eighteen hours	54°	**360°**
55 days and 1,100 minutes	55°	20°
56 days and 1,120 minutes	56°	40°
Fifty-seven days and nineteen hours	57°	60°
58 days and 1,160 minutes	58°	80°
59 days and 1,180 minutes	59°	100°
Sixty days and twenty hours	60°	120°
61 days and 1,220 minutes	61°	140°
62 days and 1,240 minutes	62°	160°
Sixty-three days and twenty-one hours	63°	180°
64 days and 1,280 minutes	64°	200°
65 days and 1,300 minutes	65°	220°
Six-six days and twenty-two hours	66°	240°
67 days and 1,340 minutes	67°	260°
68 days and 1,360 minutes	68°	280°
Sixty-nine days and twenty-three hours	69°	300°
70 days and 1,400 minutes	70°	320°
71 days and 1,420 minutes	71°	340°
Seventy-three days	72°	**360°**
One hundred forty-six days	144°	360°
Two hundred nineteen-days	216°	360°
Two hundred ninety-two days	288°	360°
Three hundred sixty-five days	360°	360°

FIVE POINT STAR

If you examine closely the chart above here, notice this pattern with the earth to move in degrees is focus around a frayation. Notice how the earth is six hours aways from its original position at the end of the year. Six hour (that is, a frayation) is surrounded by every design with the years, days, the movement of the earth, Tays, Zears et cetera. I bring this fact to your attention, because I cannot stress enough to encourage you to focus on six hours. If you focus on six hours (that is, a frayation), you will understand everything about time.

There are a lot of mystical belief about five points; here it finally has some scientific truth with time. There are only 5 points in 365 days, when the earth slide to a position at the beginning of the day, and these points are every 73 days. In other words, when the earth is positioned at 12 O'clock midnight, and one day and 20 minutes is one degree of the movement of the earth to orbit the sun, then it takes 73 days before the earth is in this position when the degree of the movement of the earth to orbit the sun is at the same time of day as the first day.

I must repeat a fact to prove another point. One day and twenty minutes can go into 365 days, 360 times, right; therefore, the earth move 1°, every 1,460 minutes, which is one day and twenty minutes. If one day and twenty minutes is 1° of the movement of the earth to orbit the sun, then 1/24 of a degree is one hour and fifty seconds. I find it very interesting that each degree of the movement of the earth to orbit the sun can be equally divided into 24 (just like a day has 24 hours). It is interested, because every function of time has a similar pattern as another function of time. I now ask you to observe when the earth has move to the position of 18° on its route to orbit the sun. Notice that it is Eighteen days and six hours. Therefore, since the new year starts at midnight, then the time of day after 18 days and six hours is 6 am in the morning. Notice, too, the location of the earth in comparison on a scale of a frayation. The position is 360°, which is the last position of a frayation. Now follow this pattern ... add 18° to this position, which is 18 days and twelve hours. The time of day is the afternoon. Now add

77

A NEW ZEAR

18° to this position, which is 54 days and 18 hours. The time of day is in the evening. Now add 18° to this position, which is 73 days. Notice how the position is exactly at the end of a frayation, and the position is a "six hours" difference, when compared to 18° prior to it. Just as the earth is six hours away at the end of the year, and a whole day after four years.

THIS IS WRONG

THANK GOD THAT life does not work as we interpret it, or there would be a catastrophe. We got the facts straight about time, but our interpretation of the fact about time is wrong, e.g., to justify our lack of understanding of the earth to position itself in a new position at the beginning of a new year, which is six hours away from the previous year, and a total of 24 hours after four years, we came up with this leap year. However, if time really works according to the leap year, life would be chaotic. Notice some of the things that could badly happen, that is, if time works like this. The extra day in the month of February would establish a total of 366 days during the leap year. It is impossible for 366 days to exist in a year. This would cause two moments in time to overlap in a year. Whatever is the position of the earth at the beginning of the year, 366 days would cause the year to end and the next year to start in a position passed the previous year. The last position cannot overlap the first position, or you have extended the year beyond 360°; in other words, this is like saying that there is 361° in a circle. Moreover, if your birthday is on February 29, then what? You only have a birthday every four years. If your birthday is after February 29, then this error is causing you to celebrate your birthday, a day late. These are just a few errors. I can go on and on, but you get my point. Let's focus more on the correct way to interpret time.

ON DEAD MAN TIME

(EPHESIANS 1:10) ... Some say that they need more time. You are giving the time that is needed to fulfill your purpose in life. It is up to you to use it wisely and efficiently

Don't say, "you don't have time"; you just wasted your time. Since GOD gave life to you, you had all the time in the world. You just did not manage to use your time properly. In other world, GOD gave time to you, and what you decide to do with your time is your decision. Therefore, people waste time; time do not waste itself. Consequently, time is your opportunity. So, use your time as an opportunity to work on your dream.

As the day before this day deliver (i.e., birth) us into this day, and this day will birth the next...

we are neither passing time, nor is it passing us. Time uses the day to bring us into the next day; therefore, we are born into the day.

You can never say that you do not have time. The ideal that you have life (i.e., alive) is proof that you have time. Now what you do with your time is in question; in other words, you have time. What are you doing

ON DEAD MAN TIME

with your time is what you are deciding to do with your time is what you are having time for ...

How precious is time, because time give the opportunity to you to do things; therefore, you have no opportunity without time; in other words, a dead man is a man without an opportunity. A dead man is the only one who don't have time

Time makes everything grow up (mentally and physical), grow old, mature... Time age everything. You can tell by observing nature. notice how the earth evolve around the sun. It causes things like food, flowers, trees, you etc. to grow, and die. Time affects everything that is composed of dirt, and it affect dirt, itself; therefore, time affect you, too, because you are made of dirt.

Today, time cause flowers to spring to blossom in its glory, and tomorrow it will wither, then fade away. Time affects us in the same manner. Since time cause every growth, time age everything. The effect of time... it is amazing what it can produce, or how it causes the result of everything. Rather you like it or not, it is a time to live, die, kill, restore, peace, war, etc. ... everything has its time. So, every dog has its day.

Time plans itself, and has plans for itself. It plans how the earth evolve around the sun, plans nature, the physical characteristic, growth, the moment, and cause the moment, and growth of everything. Man cannot determine his or her nature at birth, nor does man has the power or wisdom to change its nature. Nature is planned by time. Attempting to change nature, itself, or your nature is going against nature, which stop your natural growth ...

Notice the course of time to evolve around the sun is perfect, because it

A NEW ZEAR

produces the perfect result: flowers grow, we grow, food is produced, a new age appears, birth, death, old age etc.; now if time would alter just a little from its course of the earth to evolve around the sun, it would catastrophically change nature, i.e., crops, growth etc. would enter into utter failure. It is amazing how the arrangement of the sun, stars, moon etc. affect the earth. The earth has never deviated from its path.

DIFFERENT TIME ZONE

WE ARE AWARE of the different time zone on earth, but by my studying of time, I have noticed a completely different time system all together on other planets, which is another proof of there is no such thing as traveling through time, and neither are we nor life is traveling through time, but time is traveling through life, e.g., notice how each planet is at a different distance from the sun, which mean that it takes each planet a different amount of time to travel around the sun. Whether other planets are moving slower, faster or at the same speed as earth, each planet is at a different distance from the sun, which show that time on each planet is different (that is, time is moving slower, faster, are is longer, or shorter).

So, time is totally differently on other planets, i.e., time is recorded totally differently. Yet we see movies that show people traveling through time to other planets, when the truth is told, time on other planets has no connection to one another, but has its own independent time.

A NEW MATH

THE CONFUSION WITH time is that we are trying to make it end in the same position, which is not how time work. Now I will introduce a new operation/formula (i.e., way) for math. We want the earth to have a zero position, when there is no such thing as zero, or a negative number in time. That is, it is impossible to record someone or thing going backward in time. You can only record someone or thing with time going forward, which is one proof of the impossibility of going backward with time. Moreover, if someone gave a math problem to you, there is never a realistic moment to subtract time, e.g., measure (that is, clock) anything around you. Notice that there is no zero time, because zero time does not exist. If a stopwatch is set to zero, it has never started. Time starts after zero, and there is no negative time, because if you start a stopwatch, then stop or pause it, you can start it to continue again, but never record anything that cause the stopwatch to record to go backward in time.

TIME IS MOVING

IT IS IMPOSSIBLE to travel in time or reverse time. Since, again, time is just the indication of the position of the earth around the sun, then if the earth was reverse, and travel on its course in the opposite direction, time would not reverse. Time would still be the same, just indicating the earth is in another direction. And the earth reversing would not make life go backward, or reverse time, since the earth cannot make life reverse; moreover, since earth doesn't and cannot reverse then time cannot either, since the two are one in the same in expression. However, if the earth was to reverse its course, this would destroy the earth, because the course of the earth is what maintain the present existence and nature of the earth. It would be a different type of nature, which would not be livable.

Man tries to figure out how to travel through time, which is impossible, because life does not travel through time, time is traveling life. time is traveling through life. One proof is that since time is just a reflection of the movement of the earth on its plane around the sun, and the earth is traveling through life, then time is just a measurement of the earth to travel around the sun. Just like with every measurement! The measurement, itself, is not the real deal, but it is an indication for making a measurement of something. Since time reflects the movement of the earth on its plane around the sun, time is moving. It does not stay in the same position. If time is moving, which we all know, then it is impossible to move back in time, because time is not there anymore. Since some people do not fully understand time (for example, why there is a difference in the position of the earth at the end of the year), they try to interpret time without a complete understanding of it. How can you solve a problem, and you do not understand the problem? You just create more problems!

Like how man add a day to the year on each fourth year is throwing time off balance, and other things as well, e.g., they are trying to keep the earth in the same position at the end of the year by adding a day to the calendar on the fourth year. Notice the problem that this creates: (1). Since time reflects the movement of the earth, then man is making time sit still, when time is design to move ahead, which we all agree to this fact. Man fell to realize that the earth is not designed to be in the same position at the end of the year, which nature, itself, reveals this as a fact; since time reflects the position of the earth, then the earth is not supposed to be in the same position as time is not in the same position, but both are advancing instead of staying in the same position. The reason that I bring this up, because whenever something starts over, again, like a year, day, season et cetera, it means something. Things are happening with time without any awareness of it, which are indicating something as new and important to us. But we keep interfering instead of observing nature without interfering, and learn the truth about time, and how it works. (2). Man makes an error by adding an extra day to the calendar to try to balance something that is not out of order. It is not only throwing the interpretation of the correct time out of order and balance, it is also offsetting the correct balance, order, and birthday, e.g., if you were born after February the 28th during the year of what we call leap year, you celebrate your birthday on the wrong day, because this is the year that man say that the earth has drifted a whole day off of its course; so man add a day to position the earth back. However, whatever position of the earth for your birthday is a day

TIME IS MOVING

behind, too, according to man's erroneous theory of time. According to man's interpretation of time, the day of your birthday is supposed to be a day ahead, e.g., if you were born on February 29th, man is saying that this day should be February 28st. So, by man to add an extra day to the year, you are celebrating your birthday on the wrong day. If man would just let time takes its course without adding to time, which no man can add more time to time, but man try to anyway, you would be able to keep up with the correct time, season, years, and other periods of time, which we have fail to learn. A leap year is not the earth to drift backward a day off of its position, but it is really a year for the earth to advance a day ahead, which indicate that each four parts of the earth has now face the sun at this time of year, and now it is starting a new year with the original part of the earth to face the sun, e.g., if the position of the new year is here, then the beginning of the next new year, the position of the earth would start at a point that is six hours away from here, but advancing in six hours instead of drifting back in hours; this pattern continue four times to rotate to a new position of the earth to face the sun, and the fifth year is when the first part of the earth will face the sun, again, as was during the first years.

Time does not control life, life control time, and the two are not one in the same. For if they were one in the same, then when time stop, then life should stop. However, it was recorded in the bible that time stop, but people continue living. Even if you do not believe in the Bible, common sense tells you that if the earth was to stop, then time would stop, because time indicate the movement of earth. So, if the earth was to set still, it might change our way of living life as we know it, but it will not stop life, and everything else that exist will not stop existing. The only thing that it will affect is us. Yet if time and life was the same, then it would make everything to cease to exist, because that is the effect of life on everything. Since death is the opposite of life, time going backward would create death. I conclude that time can never travel backward.

BIBLICAL FACTS

SPEAKING OF THE Bible, it was the Bible that reveal to me regarding these new facts about time. The Bible should be study carefully, because there are other facts that are in it, which is unknown to our world, but some previous civilizations have already known these facts, e.g., Rome thought that the earth was flat, and they think that they discover the earth to be round; however, the book of Isaiah 40:22, which was recorded between 700-680 B.C., said that the earth is formed into a circle. This is one evidence of ancient civilizations, which was created a long time before the existence of Rome, to already know that the earth is round.

So, some information is really not being discover for the first time, but it is considered as lost knowledge. For every civilization has its way of technology, and some of these information (i.e., knowledge) was lost, and people are not discovering it for the first time but rediscovering lost knowledge that was discover by other ancient civilizations. So, I do not take credit for discovering this; I am just showing lost knowledge to the world that was probably discover by people before our civilization. Then, again, this could be a new discovery. There is no evidence that this was discovered by other civilizations; however, some discoveries like the discoveries by the Egyptians are kept as a secret, e.g., how did they build the pyramid, how did they learn to preserve a dead body et cetera? The evidence, which I found by other civilizations, had the correct information to discover this. Was it overlooked, or, again, just lost. So many books are destroyed with valuable information. You never know; however, thank GOD that we can now move-on and interpret time in a correct manner.

NOTES

IF YOU EXAMINE closely, that while we create a clock with the hands to move and the numbers on the clock to remain still, time does not work like that. Time advances instead of remaining in the same position when the earth reach the same location during the following year. The number on the calendar or clock is advancing, i.e., the number is shifting position at an interval of six hours. In other words, whoever design and/or created the watch, they created it with the hands to move while the numbers on the clock/watch remain still as an indication of the position to tell time; however, how the earth really work is to and on the contrary, i.e., the numbers do not indicate the positions, but the numbers move to the next position as the position of these frayation stay in the same position, and every year, the day, time, all shift to the next position of a frayation, which is an interval of six hours.

The reason why we assume time as six hours behind, because we focus on the position of the earth instead of the function of the year. The function of the year has shift six hours in advancing itself at the end of the year. Notice the earth after four years has advanced 24 hours, which is a whole day. One evidence that the earth is not drifting in position, but it is a sign of the advancement of time. If the first day of the year is on a Monday, the first day of the following year is on a Tuesday, right. Again, this is a sign of advancing (not drifting backward).

NOTES

We think that the rotation with time is over after a year, e.g., a minute is over after sixty seconds, a hour is over after sixty minutes, a day is over after twenty-four hours, and a year is over after 365 days; however, the complete cycle with time is not over at the end of a year; Years are also combine together to form a pattern.

It should be noted that it is a coincidence, when it occurs maybe sometime, or sporadic; nevertheless, when it is a normal repeated pattern, which is happening constantly with the same periodic interval, it is not a coincidence, but a pattern. Therefore, you cannot find this as a coincidence that there are 1,460 six-hours in a year, as there are 1,460 days in four years, and 1,460 years for the earth to form its pattern. Moreover, there are 365 frayations in ¼ of year, which is equivalent to 91.25 days, and reflects the earth to rotate 90 degrees around the sun. Notice what 365 days, and 365 frayations have in common. They both have rotated ¼ of 360° of its circle; moreover, 1,460 frayations is the total for a complete year, and 1,460 days are the total for the earth to rotate to start the new year in the same position. Moreover, 365 years reflect the same pattern as 365 days, and 365 frayations; consequently, 1,460 years reflect the same pattern as 1460 days, and 1,460 frayations.

Since the sun is not moving, but the earth is moving, then the sun is not setting, or rising; therefore, we should use terms to indicate the earth in the morning and in the evening instead of the sun. Since the earth is moving, notice that the sun is seen in the east in the morning, then in the west in the evening. This is concrete evidence that the earth is turning counterclockwise during the day.

ENDNOTES

1. Frayation means six hours, which is the foundational position to start and end each day, week, month, year, Tay, Deason, Zear etc.
2. A frayation is six hours, which is the foundational position to start and end each day, week, month, year, Tay, Deason, Zear etc.
3. A comyetion is 10,220 years, which is a completion of the cycle of time.
4. 10,220 years
5. A frayation is six hours, which is the foundational position to start and end each day, week, month, year, Tay, Deason, Zear etc.
6. A Tay is four years.
7. I repeat, a Tay is four years.
8. 1,460 years is a Zear.
9. 365 years is a Deason.
10. I repeat, a Zear is 1,460.
11. 10,220 years is a Comyetion – The earth start a new year in its original position as the first day

www.ingramcontent.com/pod-product-compliance
Lightning Source LLC
Chambersburg PA
CBHW070309230526
45470CB00002B/789